Biodiversity Conservation – Challenges for the Future

Edited By

Laladhas K.P., Oommen V. Oommen

Kerala State Biodiversity Board
L-14, Jai Nagar
Medical College P.O.
Thiruvananthapuram-695 011
India

&

Sudhakaran P.R.

Kerala Biotechnology Commission
Kerala State Council for Science
Technology and Environment
Shastra Bhavan, Pattom
Thiruvananthapuram - 695 004
India

CONTENTS

Foreword *i*

Preface *iii*

List of Contributors *v*

CHAPTERS

SECTION 1: SUSTAINABLE UTILIZATION, MANAGEMENT AND CONSERVATION OF BIORESOURCES

1. **The Diversity and the Ecological Status of Snakes in Kerala Region - A View to Prioritize Conservation Measures** 3
Dileep Kumar R., Anaswara Krishnan S., Laladhas K. P. and Oommen V. Oommen

2. **Conservation of Medicinal Plant Resources through Community Born Biodiversity Management Committee, Kerala, India** 27
Laladhas K. P., Preetha N., Baijulal B. and Oommen V. Oommen

3. **The State of Frogs in India: A Case for Discontinuation of Use of Frogs in Dissections and Experiments** 41
Mohammad A. Akbarsha

4. **The Role of People's Participation in the Conservation and Management of Biodiversity and Associated Traditional Knowledge** 53
Rajasekharan S.

5. **Current and Likely Trends in Biodiversity Conservation** 65
Rajmohan K.

6. **Natural Bioresources of North Eastern Region of India - Its Conservation and Proper Utilization** 71
Unni B.G., Bharadwaz M., Kalita M., Dey T., Gogoi K., Kalita J., Islam S.J. and Dutta P.

7. **Sacred Groves of Himachal Pradesh: a Traditional Community Practice of Biodiversity Conservation and Testimony of Religious Beliefs** 89
Vandana Thaplyal and Ashok Verma

8. **Population Density and Conservation Status of Indian Gaur (*Bos gaurus gaurus*) in Trishna Wildlife Sanctuary, Tripura, Northeast India** 97
Mary P. P., Debbarma J., Kumar Awadhesh and Gupta A. K.

SECTION 2: TRADITIONAL AND INDIGENOUS SEEDS AND BREEDS FOR THE FUTURE

9. **Therapeutic Value of Milk of Vechur Cow, the Indigenous Breed of Kerala Listed as Critical by FAO** 103
Mohankumar Chinnamma, Anisha Shashidharan and Salini Bhasker

10. **An Account on Some Endemic Seeds of Southern Western Ghats** 113
Anilkumar C., Chitra C. R., Bindu S., Maya P. V., Prajith T. M. and Joemon J.

SECTION 3: SUSTAINABLE FOOD SECURITY

11. **Non Wood Forest Biodiversity for Food Security- Rattans a Case Study** 121
Manohara T. N.

12. **Collective Action on Improving Environmental and Economic Performance of Vegetable Production: Exploring Pesticides Safety in India** 127
Sreejith Aravindakshan and Sherief A. K.

13. **Documentation and Farmer Perception on Biodiversity in Rice Fields of Palakkad District** 137
Smitha K.P. and Anilkumar A.

SECTION 4: ACCESS AND BENEFIT SHARING OF BIORESOURCES

14. **An Initiative towards Achieving the 'Access and Benefit Sharing' Provisions of Biodiversity Act 2002 and Rules 2004, India** 151
Sudipta Mukherjee, Soumyendra Nath Ghosh and Debal Ray

SECTION 5: GREEN ECONOMICS

15. **Actionable Innovative Financial and Economic Instruments for Biodiversity Conservation in India** 157
Damodaran A.

16. **Green Economics: It's Impact on Environment and Biodiversity** 175
Jayamani C. V.

17. **Biodiversity Valuation of Marine Ecosystem of the South-West Coast of India with Special Reference to Kerala Coast** 183
Joshi K. K., Narayanakumar R., Ranimary George, Molly Varghese, Jasmine S., Srinath K. R., Saravanan S., Somy Kuriakose, Geetha Antony, Geetha P. M. and Gimy George

SECTION 6: BIODIVERSITY – THE WEB OF LIFE

18. **Survey and Assessment of the Status, Issues & Threats to Marine Turtles Along the Kerala Coast** 195
Renjan Mathew Varghese

19. **Synchronous and Rhythmic Light Display by a Panoramic Congregation of Fireflies at Varagaliar, Anamalai Tiger Reserve** 203
Mathish Nambiar-Veetil, Selvaraj P., Blessan Santhosh George, Ganesan M., Raghunath T. P. and Krishnakumar N.

20. **Identification, Purification and Mass Spectrometric Characterization of Two Novel Cyclic Peptides from *Clitoria ternatea*** 209
Sudarslal S., Balaram P. and Laladhas K. P.

21. **Satellite Based Spectral Signatures of Coastal Wetlands and Mangrove Vegetation of Kasaragod, India** 217
Mohammed-Aslam M. A.

22. **Diversity of Stingless Bee Flora in Kerala** 221
Devanesan S., Premila K. S. and Shailaja K. K.

23. **Seasonal Variations Influencing the Bryophyte Diversity of Monoculture Plantations in the Southern Western Ghats** 229
Kariyappa K. C. and Daniels A. E. D.

24. **Macroinvertebrates Associated with Aquatic Macrophytes *Monochoria vaginalis* and *Eichhornia crassipes* in Ponnani Kole** 235
Sanila V. K. and Razia Beevi M.

25. **Micropropagation of an Exquisite Orchid, *Acampe rigida* (Buch. – Ham. Ex J. E. Sm.) P.F. Hunt., Through Embryo Culture** 241
Gangaprasad A., Anjusha S. and Sibin N. P.

SECTION 7: BIOIDIVERSITY CONSERVATION- CHALLENGES FOR FUTURE AND WAY FORWARD

26. **Biodiversity Conservation - Challenges for Future and Way Forward** 249
Preetha N., Laladhas K. P. and Oommen V. Oommen

Authors Index 257

Subject Index 259

FOREWORD

It is not often that meetings and conferences discuss and come up with specific ideas to make science inform policy and policy directions suggested, based on sound science. The National Conference on Biodiversity, organized by the Kerala State Biodiversity Board in association with a range of organizations, including the National Biodiversity Authority, Government of India during December 2012 has precisely done this.

The present volume of material provides readers with a rainbow of issues related to knowing and managing India's biological diversity including the status of these resources while the recommendations indicate the need for appropriate action plans and policy directions for India's biodiversity management. This compilation comes at a time when the world is watching India, patiently, to come up with its guidance to deal with biodiversity related actions as President of the 194 country strong Conference of Parties to the Convention on Biological Diversity.

Biodiversity is life! It has been amply demonstrated that natural resources offer the best insurance for local people in dealing with adaptation and livelihood securities. While scientists have continued to publish volumes on knowing the diversity and suggested the ways to manage it, the results of their suggestions have had limited impact on countries like India. One of the reasons for this is the wait and watch policy of Government agencies, while addressing and making use of local opportunities that had taken a back seat. With India's extensive expertise on various areas of biodiversity, time has come for us to re-think the need to re-connect to local action and ensure that work by academic institutions and non-governmental organizations do influence and impact Government policies. The national biodiversity management options also need a complete over-haul during this decade of biodiversity (2011-2020). It is important for all stakeholders to understand that waiting for Government to act is not the only option. Publications, such as these reports, are timely to provide a beacon of hope for the future of biodiversity in India where the spread and depth of action and options need to be measured and used.

Personally, I feel very encouraged that the Kerala State Biodiversity Board, under the guidance of Honourable Chief Minister of Kerala Shri Oommen Chandy, has taken this initiative of organizing the national meeting on biodiversity. The premise of establishing the State Biodiversity Boards under the Biological Diversity Act in India is to provide the option for decentralized implementation of the Act and the State Boards need to precisely engage in such activities such as this. It is important, however, that this kind of dialogues and meetings as well as compilations find a regular place in the national events calendar. In the absence of this, such outcomes and their relevance will fade away.

It is also heartening to write this Foreword since several of the suggestions made in the recommendations section have been followed up by the National Biodiversity Authority during my tenure. The proposal to establish the National Biodiversity Grid has been approved by the Office of Principal Science Advisor to the Prime Minister of India, announcement to launch the National Innovation Council on Biodiversity was made in January 2014 and a range of actions to raise awareness and capacities on issues of biodiversity implemented.

We will certainly fail in our duties, if we do not recognize the unique and leadership role taken by the Kerala State Biodiversity Board in implementing the Biological Diversity Act and in providing a leadership role in the country on issues of establishing and training the Biodiversity Management Committees, preparing the People's Biodiversity Registers, empowering communities to manage livelihoods using bioresources and most importantly, helping the State Government in making appropriate public policies on biodiversity.

I hope that the present publication will not only help several of us understand the Current Challenges to deal with biodiversity in India but such a publication should also reach Schools, Universities and research institutions so that we have future corps of followers to move the conservation and development agenda in India that can never be delivered by the Government alone.

Dr. Balakrishna Pisupati
Former Chairman
National Biodiversity Authority (NBA)
Chennai, Tamil Nadu
India

PREFACE

The National Biodiversity Congress held in 2012 inspired the compilation of this book in the present form. The first of its meetings was held in the South Indian city of Thiruvananthapuram jointly by the National Biodiversity Authority, India and Kerala State Biodiversity Board, both organisations involved in the conservation of biodiversity of the country as envisaged in the Biological Diversity Act 2002. The Congress was attended by political managers, academicians, scientists and officials of the various State Biodiversity Boards across the country and views exchanged on different aspects of biodiversity conservation, its challenges and the way forward. During his key note address to the congress, Dr. R Chidambaram, Principal Science Advisor to Government of India, confirmed the establishment of a National Biodiversity Grid to facilitate and support the work in collating, analyzing and synthesizing nation-wide biodiversity data and information for conservation and sustainable management of biodiversity of the country and also to assist appropriate policy making. Financing biodiversity was identified as vital for its conservation and management and mobilizing adequate finance is acknowledged as a huge challenge during the ensuing discussions. Access and Benefit Sharing (ABS) was identified as a key emerging internal financial option. Mainstreaming of BD issues-linking to planning process of Local bodies is an important prerequisite in strengthening the institutional capacity for good local governance. Conservation history is being rewritten by cumulative conservation efforts and this book is an effort to compile the main initiatives on biodiversity governance globally as well as a compilation of selected presented in the various scientific sessions during the congress.

Laladhas K.P., Oommen V. Oommen

Kerala State Biodiversity Board
L-14, Jai Nagar
Medical College P.O.
Thiruvananthapuram-695 011
India

&

Sudhakaran P.R.

Kerala Biotechnology Commission
Kerala State Council for Science
Technology and Environment
Shastra Bhavan, Pattom
Thiruvananthapuram - 695 004
India

List of Contributors

Anaswara Krishnan S.	Department of Zoology, University of Kerala, Kerala, India
Anilkumar A.	Dept of Agrl.Extension, College of Agriculture, Vellayani, Thiruvananthapuram, Kerala, India
Anilkumar C.	Jawaharlal Nehru Tropical Botanic Garden and Research Institute, Palode, Thiruvananthapuram, Kerala, India
Anisha Shashidharan	SCMS Institute of Bioscience and Biotechnology Research and Development, Management House, South Kalamassery, Cochin, Kerala, India
Anjusha S.	Plant Tissue Culture and Molecular Biology Lab, Department of Botany, University of Kerala, Kariyavattom, Thiruvananthapuram, Kerala, India
Ashok Verma	WWF-India, Field Office, Bishop Cotton School, Shimla, Himachal Pradesh, India
Baijulal B.	Kerala State Biodiversity Board, L-14, Jai Nagar, Medical College P.O, Thiruvananthapuram, Kerala, India
Balaram P.	Molecular Biophysics Unit, Indian Institute of Science, Bangalore, Karnataka, India
Bharadwaz M.	Biotechnology Division, CSIR-North East Institute of Science and Technology, Jorhat, Assam, India
Bindu S.	Jawaharlal Nehru Tropical Botanic Garden and Research Institute, Palode, Thiruvananthapuram, Kerala, India
Blessan Santhosh George	Institute of Forest Genetics and Tree Breeding, Forest Campus, R. S. Puram, Coimbatore, India
Chitra C. R.	Jawaharlal Nehru Tropical Botanic Garden and Research Institute, Palode, Thiruvananthapuram, Kerala, India
Damodaran A.	Indian Institute of Management, Bangalore, Karnataka, India
Daniels A. E. D.	Bryology Laboratory, Department of Botany & Research Centre, Scott Christian College, Nagercoil, Tamil Nadu, India
David J. Gower	Department of Zoology, The Natural History Museum, London, SW7 5BD, UK
Debal Ray	West Bengal Biodiversity Board, Poura Bhawan, 4th Floor, FD-415A, Sector – III, Salt Lake City, Kolkata, India
Debbarma J.	North Eastern Regional Institute of Science and Technology (NERIST) (Deemed University), Arunachal Pradesh, India
Devanesan S.	Honey bees and Pollinators, Vellayani Centre, Department of Agricultural Entomology, College of Agriculture, Kerala Agricultural

University, Vellayani, Kerala, India

Dey T. Biotechnology Division, CSIR-North East Institute of Science and Technology, Jorhat, Assam, India

Dileepkumar R. Centre for Venom informatics, University of Kerala, Kerala

Dutta P. Biotechnology Division, CSIR-North East Institute of Science and Technology, Jorhat, Assam, India

Ganesan M. Institute of Forest Genetics and Tree Breeding, Forest Campus, R. S. Puram, Coimbatore, India

Gangaprasad A. Plant Tissue Culture and Molecular Biology Lab, Department of Botany, University of Kerala, Kariyavattom, Thiruvananthapuram, Kerala, India

Geetha Antony Marine Biodiversity Division, Central Marine Fisheries Research Institute, Kochi, Kerala, India

Geetha P. M. Marine Biodiversity Division, Central Marine Fisheries Research Institute, Kochi, Kerala, India

Gimy George Marine Biodiversity Division, Central Marine Fisheries Research Institute, Kochi, Kerala, India

Gogoi K. Biotechnology Division, CSIR-North East Institute of Science and Technology, Jorhat, Assam, India

Gupta A. K. Dept. of Environment & Forest, Govt. of Tripura, P. O. Kunjaban, Agartala, Tripura (W), India

Islam S. J. Biotechnology Division, CSIR-North East Institute of Science and Technology, Jorhat, Assam, India

Jacob J. Jawaharlal Nehru Tropical Botanic Garden and Research Institute, Palode, Thiruvananthapuram, Kerala, India

Jasmine S. Marine Biodiversity Division, Central Marine Fisheries Research Institute, Kochi, Kerala, India

Jayamani C. V. School of Management Studies, CUSAT, Kochi, Kerala, India

Joshi K. K. Marine Biodiversity Division, Central Marine Fisheries Research Institute, Kochi, Kerala, India

Kalita J. Biotechnology Division, CSIR-North East Institute of Science and Technology, Jorhat, Assam, India

Kalita M. Biotechnology Division, CSIR-North East Institute of Science and Technology, Jorhat, Assam, India

Kariyappa K. C. Bryology Laboratory, Department of Botany & Research Centre, Scott Christian College, Nagercoil, Tamil Nadu, India

Krishnakumar N. Institute of Forest Genetics and Tree Breeding, Forest Campus, R. S.

Puram, Coimbatore, India

Kumar Awadhesh North Eastern Regional Institute of Science and Technology (NERIST) (Deemed University), Arunachal Pradesh, India

Laladhas K. P. Molecular Biophysics Unit, Indian Institute of Science, Bangalore, Karnataka, India and Biomolecular Research Unit, Department of Zoology, St. Stephen's College, Pathanapuram, Kollam, Kerala, India and Kerala State Biodiversity Board, L-14, Jai Nagar, Medical College P.O., Thiruvananthapuram, Kerala, India

Lekha Divya Department of Animal Sciences, School of Biological Science, Central University of Kerala, Riverside campus, Padanakadu, Kasaragod, India

Manohara T. N. Rain Forest Research Institute, Jorhat, Assam, India

Mark Wilkinson Department of Zoology, The Natural History Museum, London, SW7 5BD, UK

Mary P. P. North Eastern Regional Institute of Science and Technology (NERIST) (Deemed University), Arunachal Pradesh, India

Mathish Nambiar-Veetil Institute of Forest Genetics and Tree Breeding, Forest Campus, R. S. Puram, Coimbatore, India

Maya P. V. Jawaharlal Nehru Tropical Botanic Garden and Research Institute, Palode, Thiruvananthapuram, Kerala, India

Mohammad A. Akbarsha Gandhi-Gruber-Doerenkamp Chair, Mahatma Gandhi-Doerenkamp Center (MGDC) for Use of Animals in Life Science Education, Bharathidasan University, Tiruchirappalli,Tamilnadu, India

Mohammed-Aslam M. A. Department of Geology, Central University of Karnataka, Gulbarga, Karnataka, India

Mohankumar Chinnamma, SCMS Institute of Bioscience and Biotechnology Research and Development, Management House, South Kalamassery, Cochin, Kerala, India

Molly Varghese Marine Biodiversity Division, Central Marine Fisheries Research Institute, Kochi, Kerala, India

Narayanakumar R. Marine Biodiversity Division, Central Marine Fisheries Research Institute, Kochi, Kerala, India

Oommen V. Oommen Kerala State Biodiversity Board, L- 14, Jai Nagar, Medical College P O, Thiruvananthapuram, Kerala, India

Prajith T. M. Jawaharlal Nehru Tropical Botanic Garden and Research Institute, Palode, Thiruvananthapuram, Kerala, India

Preetha N. Kerala State Biodiversity Board, L-14, Jai Nagar, Medical College P.O., Thiruvananthapuram, Kerala, India

Premila K. S.	Honey bees and Pollinators, Vellayani Centre, Department of Agricultural Entomology, College of Agriculture, Kerala Agricultural University, Vellayani, Kerala, India
Raghunath T. P.	Institute of Forest Genetics and Tree Breeding, Forest Campus, R. S. Puram, Coimbatore, India
Rajasekharan S.	Division of Ethnomedicine and Ethnopharmacology, Jawaharlal Nehru Tropical Botanic Garden and Research Institute (JNTBGRI), Palode, Thiruvananthapuram, Kerala, India
Rajmohan K.	Kerala Agricultural University, College of Agriculture, Vellayani, Thiruvananthapuram, Kerala, India
Ramachandran Kotharambath	Department of Zoology, Government College Chittur, Palakkad, Kerala, India
Ranimary George	Marine Biodiversity Division, Central Marine Fisheries Research Institute, Kochi, Kerala, India
Razia Beevi M.	Department of Aquaculture and Fishery Microbiology, M.E.S. Ponnani College, Ponnani, Malappuram Dt., Kerala, India
Renjan Mathew Varghese	WWF-India, Kerala State Office, Vanchiyoor P.O, Thiruvananthapuram, Kerala, India
Ronald A. Nussbaum	Museum of Zoology and Department of Ecology and Evolutionary Biology, University of Michigan, Ann Arbor, MI 48109-1079, USA
Salini Bhasker	SCMS Institute of Bioscience and Biotechnology Research and Development, Management House, South Kalamassery, Cochin, Kerala, India
Sanil George	Molecular Medicine Laboratory, Rajiv Gandhi Centre for Biotechnology, Thiruvananthapuram, Kerala, India
Sanila V. K.	Department of Aquaculture and Fishery Microbiology, M.E.S. Ponnani College, Ponnani, Malappuram Dt., Kerala, India
Saravanan S.	Marine Biodiversity Division, Central Marine Fisheries Research Institute, Kochi, Kerala, India
Selvaraj P.	Institute of Forest Genetics and Tree Breeding, Forest Campus, R. S. Puram, Coimbatore, India
Shailaja K. K.	Honey bees and Pollinators, Vellayani Centre, Department of Agricultural Entomology, College of Agriculture, Kerala Agricultural University, Vellayani, Kerala, India
Sherief A. K	Department of Extension, Kerala Agricultural University, India
Sibin N. P.	
Smitha K. P.	Community Agrobiodiversity Centre, M.S.Swaminathan Research Foundation, Puthurvayal P.O, Kalpetta, Wayanad, Kerala, India

Somy Kuriakose	Marine Biodiversity Division, Central Marine Fisheries Research Institute, Kochi, Kerala, India
Soumyendra Nath Ghosh	West Bengal Biodiversity Board, Poura Bhawan, 4th Floor, FD-415A, Sector – III, Salt Lake City, Kolkata, India
Sreejith Aravindakshan	Farming Systems Ecology, Wageningen UR, the Netherlands, SUTROFOR Erasmus Mundus Fellow, LIFE, University of Copenhagen, Denmark (previously)
Srinath K. R.	Marine Biodiversity Division, Central Marine Fisheries Research Institute, Kochi, Kerala, India
Sudarslal S.	Molecular Biophysics Unit, Indian Institute of Science, Bangalore, Karnataka, India and Amrita School of Biotechnology, Amrita Vishwa Vidyapeetham, Kollam, Kerala, India
Sudipta Mukherjee	West Bengal Biodiversity Board, Poura Bhawan, 4th Floor, FD-415A, Sector – III, Salt Lake City, Kolkata, India
Unni B.G.	Biotechnology Division, CSIR-North East Institute of Science and Technology, Jorhat, Assam, India
Vandana Thaplyal	WWF-India, Field Office, Bishop Cotton School, Shimla, Himachal Pradesh, India

CHAPTER 1

The Diversity and the Ecological Status of Snakes in Kerala Region-A View to Prioritize Conservation Measures

Dileep Kumar R.[1,*], Anaswara Krishnan S.[2], Laladhas K. P.[3] and Oommen V. Oommen[3]

[1]Centre for Venom informatics, University of Kerala, Kerala-695581; [2]Department of Zoology, University of Kerala, Kerala-695581 and [3]Kerala State Biodiversity Board, L-14, Jai Nagar, Medical College P O, Thiruvananthapuram, Kerala-695 024

Abstract: Snakes are ecologically significant reptilian groups, but very little measures have been taken towards exploring their diversity. Kerala, located at the southern tip of the Indian subcontinent has a countable richness in snake species, approximately 104 species including endangered and endemic species. Even though there is an abundance of species diversity, a critical turn down in the population of many snake species has been encountered and it is necessary to tackle this threat of decline in population. Snakes as a whole have very little commercial value as compared with other group of animals. The various ways in which they are exploited tend to be limited to specific regions or to a few species. In addition to exploitation, habitat destruction and pollution are the most serious threats affecting their diversity. The physical structure and thermal properties of the inhabiting sites are crucial to these ectothermic creatures; and so the continuous decline in the quality of habitat is the fundamental cause that makes these species vulnerable. Extended analysis of species-habitat associationship is crucial for effectual and targeted conservation planning and management. The critical limitation found in conservation planning is the nonavailability of statistics on population size of individual species and their habitat requirements. This chapter provides an overview of the distribution and the ecological status of snake species in Kerala and also highlights the threats and possible conservation strategies for protecting their diversity.

Keywords: Snakes, diversity, anthropogenic, ecology, ecosystem, ectothermic, hotspot, parthenogenic, predator, prey, conservation, Western Ghats.

1. INTRODUCTION

Among all the animal species, snakes are the most widely known and they evoke profound sense of wonder and fear among human beings. This fear has prevailed

***Corresponding author Dileepkumar R.:** Centre for Venom informatics, University of Kerala, Kerala-695581; Tel: +919447830909; Fax: +914712308759; E-mail: dileepkamukumpuzha@gmail.com

Laladhas K.P., Oommen V. Oommen and Sudhakaran P.R. (Eds.)

because of its deadly venom even though many snake species are nonvenomous. In Indian subcontinent, relationship with snakes' traces back to the dawn of history and because of deep rooted myths, they are even worshipped till date. As per taxonomical classification, snakes belong to order *Squamata* and sub-order *Serpentes*. Like all reptiles, snakes are also cold blooded species as they regulate their body temperature to that of the surroundings. There are more than 2500 different species of snakes known to science. All have certain common anatomical features which can be summarized as-elongated and scaly body, a forked tongue, and lack of eyelids, limbs and external ears. A few of the snake species give birth to young ones while most of them are egg laying ones; rarely parthenogenic. Without exception, snakes are carnivores by nature and commonly consumed organisms include termites, fishes, rodents, birds, frogs, small deer and other reptiles. Skull of snakes have a specialized bone, the quadrate bones which enables them to move their upper jaws relative to the braincase rendering greater flexibility to the jaws. This flexible property of the jaws makes their bites much more powerful in relation to their size. Also they can completely dislocate their lower jaws, which allows them to swallow their prey whole.

2. SNAKES OF KERALA-AN OVERVIEW

The tropical region provides suitable habitat for the survival of a range of snake varieties including poisonous to non poisonous, large to small, desert snakes to sea snakes *etc*. Greater species richness is primarily due7 to the relatively uniform warm and humid climatic conditions in the region. Western Ghats, India, is one of the places where the highest species richness among reptilian species is recorded. Kerala state favours the inhabitation of diverse snake species, many of them being endemic to the region (Fig. **1**). Species categorized as threatened and data deficient has also been reported from Kerala. From the so far recorded data of the snake species in Kerala region, around 104 species in 9 families are identified and reported (Figs. **2.1** to **2.29**). Common Cobra (*Naja naja*), Common Krait (*Bunguraus caeruleus*), Russell's viper (*Daboia russelii*) *etc.* are the most common venomous snakes found throughout Kerala and most of the snake bite accidents are due to them. King cobra (*Ophiophagus hannah*) and Hump Nosed Pit Viper (*Hypnale hypnale*) are the other venomous species. The longest venomous snake, King Cobra, is rarely encountered in this range. The Kerala mud snake (*Dieurostus dussumierii*), the water snake which is endemic to Kerala has remained relatively unknown to science, because of its restricted distribution.

Figure 1: Map of Kerala showing Western Ghats region.

The unavailability of data of the population size of each species hinders the categorization of exact ecological status of the snake species and also the snake-habitat relationships. This chapter lists a collated data on the whole snake fauna spotted in Kerala with its current population status (Table **1**) and also a comprehensive overview of potential threats along with possible strategies for effective conservation. The data are derived primarily from literature sources, together with some observations by the authors.

Table 1: List of snake species inhabiting the Kerala region with its current status

Sl.No	Scientific Name	Common Name	Category Population and Current Status*
Family: Typhlopidae Merrem, 1820			
1	***Ramphotyphlops braminus* (Daudin, 1803)**	Common Blind Snake, Brahminy Worm Snake	Non-venomous Common, Not assessed
2	***Typhlops porrectus* Stoliczka, 1871**	Slender Worm Snake	Non-venomous Common, Not assessed
3	***Typhlops thurstoni* Boettger, 1890**	Thurston's Worm Snake	Non-venomous Data deficient
4	***Typhlops beddomei* Boulenger, 1890**	Beddome's Worm Snake	Non-venomous Rare, Not assessed
5	***Typhlops tindalli* Smith, 1943**	Malabar Worm Snake, Tindall's Worm Snake	Non-venomous Data deficient
6	***Grypotyphlops acutus* (Dumeril & Bibron, 1844)**	Beaked Worm Snake	Non-venomous Common, Least Concern.
Family: Uropeltidae Muller,1832			
7	***Melanophidium punctatum* Beddome,1871**	Pied-belly Shieldtail, Beddome's Black Earth Snake	Non-venomous Common, Least Concern
8	***Melanophidium bilineatum* Beddome 1870**	Yellow-striped Shieldtail, Two-lined Black earth Snake	Non-venomous Rare, **Vulnerable**
9	***Melanophidium wynaudense* (Beddome,1863)**	Wynad Shieldtail, Indian Black Earth Snake	Non-venomous Common, Least Concern
10	***Platyplectrurus trilineatus* (Beddome, 1867)**	Three-lined Shieldtail, Lined Thorntail Snake	Non-venomous Data deficient
11	***Platyplectrurus madurensis* Beddome 1877**	Madurai Shieldtail, Travancore Hills Thorntail Snake	Non-venomous Rare, **Endangered**
12	***Teretrurus sanguineus* (Beddome, 1867)**	Purple-red Earth Snake	Non-venomous Common, Least Concern
13	***Plectrurus guentheri* (Beddome, 1863)**	Gunther's Burrowing Snake	Non-venomous Data deficient
14	***Plectrurus perrotetii* Dumeril, 1851**	Nilgiri Burrowing Snake, Perrotet's Shieldtail, Perrotet's Burrowing Snake	Non-venomous Common, Least Concern

Table 1: contd...

15	***Plectrurus aureus* Beddome, 1880**	Kerala Burrowing Snake	Non-venomous Data deficient
16	***Brachyophidium rhodogaster* Wall, 1921**	Red-bellied Shieldtail, Palni Mountain Burrowing Snake	Non-venomous Common, Least Concern
17	***Uropeltis nitida* (Beddome, 1878)**	Cochin Shieldtail, Southern Earth Snake	Non-venomous Data deficient
18	***Uropeltis ocellata* (Beddome, 1863)**	Ocellated Earth Snake, Nilgiri Uropeltis, Ocellate Shieldtail	Non-venomous Common, Least Concern
19	***Uropeltis woodmasoni* (Theobald, 1876)**	Woodmason's Earth Snake, Black-bellied Shieldtail, Palni Uropeltis	Non-venomous Common, Least Concern
20	***Uropeltis ceylanica* Cuvier, 1829**	Kerala Shieldtail, Cuvier's Shieldtail	Non-venomous Common, Least Concern
21	***Uropeltis arcticeps* Gunther, 1875**	Madura Earth Snake, Gunther's Shieldtail, Thirunelveli Uropeltis	Non-venomous Common, Least Concern
22	***Uropeltis phipsonii Mason, 1888***	Phipson's Shieldtail	Non-venomous Common, **Vulnerable**
23	***Uropeltis rubrolineata* (Gunther, 1875)**	Red-lined Earth Snake, Red-striped Uropelt	Non-venomous Common, Least Concern
24	***Uropeltis myhendrae* (Beddome, 1886)**	Barred Shieldtail, Boulenger's Earth Snake	Non-venomous Data deficient
25	***Uropeltis rubromaculata* (Beddome, 1867)**	Red-spotted Shieldtail, Beddome's Uropeltis, Red-spotted Earth Snake	Non-venomous Common, Least Concern
26	***Uropeltis maculata* (Beddome, 1867)**	Orange-spotted Uropeltis, Spotted Earth Snake	Non-venomous, Data deficient
27	***Uropeltis pulneyensis* (Beddome, 1863)**	Palni Shieldtail, Indian Earth Snake	Non-venomous Common, Least Concern
28	***Uropeltis ellioti* (Gray, 1858)**	Elliot's Uropelt, Elliot's Shieldtail, Elliot's Earth Snake	Non-venomous Common, Least Concern
29	***Uropeltis beddomei* (Gunther, 1862)**	Beddome's Uropelt, Beddome's Earth Snake	Non-venomous Data deficient

Table 1: contd…

30	***Uropeltis macrorhyncha* (Beddome, 1877)**	Anaimalai Earth Snake, Anaimalai Earth Snake	Non-venomous Data deficient
31	***Rhinophis sanguineus* (Beddome, 1863)**	Red-bellied Uropelt, Red-bellied Shieldtail, Salty Earth Snake	Non-venomous Common, Least Concern
32	***Rhinophis fergusonianus* Boulenger, 1896**	Cardamom Shieldtail, Cardamom Hill's Earth Snake	Non-venomous Data deficient
33	***Rhinophis travancoricus* Boulenger, 1893**	Travancore Shieldtail, Travancore Earth Snake, Tamil Nadu Earth Snake	Non-venomous Rare, **Endangered**
Family: Pythonidae Bonaparte, 1845			
34	***Python molurus* (Linnaeus, 1758)**	Indian Rock Python, Asiatic Rock Python, Burmese Python, Indian Python, Tiger Python	Non-venomous Common, **Near Threatened**
Family: Boidae Gray, 1825			
35	***Gongylophis conicus* (Schneider, 1801)** ***Gongylophis conicus conicus* (Schneider, 1801)**	Common Sand Boa	Non-venomous Common, Not assessed
36	***Eryx johnii* (Russell, 1801)** ***Eryx johnii johnii* (Russell, 1801)**	Red Sand Boa, Eastern Red Sand Boa	Non-venomous Common, Not assessed
37	***Eryx whitakeri* Das, 1991**	Whitaker's Boa	Non-venomous Common, Not assessed
Family: Acrochordidae Bonaparte, 1831			
38	***Acrochordus granulatus* (Schneider, 1799)**	File Snake, Indian Wart Snake, Elephant Trunk Snake	Non-venomous Common, Least Concern
Family: Colubridae Oppel, 1811			
39	***Coelognathus helena* (Daudin, 1803)** ***Coelognathus helena helena* (Daudin, 1803)** ***Coelognathus helena monticollaris* (Schulz, 1992)**	Trinket snake, Common Trinket Snake, Montane Trinket snake	Non-venomous Common, Not assessed
40	***Ptyas mucosa* (Linnaeus, 1758)** ***Ptyas mucosa mucosa* (Linnaeus, 1758)**	Indian Rat Snake	Non-venomous Common, Not assessed

Table 1: contd…

41	***Argyrogena fasciolata* (Shaw, 1802)**	Banded Racer	Non-venomous Unccmmon, Not assessed
42	***Liopeltis calamaria* (Gunther, 1858)**	Faintly Stripe-necked Snake, Lesser Stripe-necked Snake	Non-venomous Uncommon, Not assessed
43	***Oligodon venustus* (Jerdon, 1853)**	Black-spotted Kukri Snake, Jerdon's Kukri Snake	Non-venomous Common, Least Concern
44	***Oligodon travancoricus* Beddome, 1877**	Travancore Kukri Snake	Non-venomous Data deficient
45	***Oligodon arnensis* (Shaw, 1802)**	Common Kukri Snake	Non-venomous Common, Not assessed
46	***Oligodon taeniolatus* (Jerdon, 1853)** ***Oligodon taeniolatus fasciatus* (Gunther, 1864)**	Russell's Kukri Snake, Streaked Kukri Snake	Non-venomous Common, Least Concern
47	***Oligodon affinis* Gunther, 1862**	Western Kukri Snake, Malabar Brown Kukri Snake	Non-venomous Common, Least Concern
48	***Oligodon brevicauda* (Gunther, 1862)**	Striped Kukri Snake, Short-tailed Kukri Snake	Non-venomous Rare, **Vulnerable**
49	***Dendrelaphis tristis* (Daudin, 1803)**	Common Indian Bronze-back	Non-venomous Common, Not assessed
50	***Dendrelaphis grandoculis* (Boulenger, 1890)**	Large-eyed Bronzeback, Southern Bronzeback	Non-venomous Common, Least Concern
51	***Dendrelaphis bifrenalis* (Boulenger, 1890)**	Black-striped Bronzeback, Boulenger's Bronzeback, Boulenger's Bronzeback, Travancore Bronze-brown Snake	Non-venomous Common, Least Concern
52	***Dendrelaphis pictus (Gmelin, 1789)*** ***Dendrelaphis pictus pictus (Gmelin, 1789)*** ***Dendrelaphis pictus andamanensis (Gmelin, 1789)***	Painted Bronzeback Tree Snake, Common Painted Bronzeback Tree Snake, Andaman Painted Bronzeback Tree Snake	Non-venomous Common, Not assessed
53	***Dendrelaphis caudolineolatus* (Gunther, 1869)**	Striped-tailed Bronze-back	Non-venomous Rare, Not assessed
54	***Chrysopelea ornata* (Shaw, 1802)**	Ornate Flying Snake, Indian Ornate Flying Snake	Mildly venomous Uncommon, Not

Table 1: contd…

	***Chrysopelea ornata* (Shaw, 1802)** ***Chrysopelea ornate ornatissima* Werner,1925**	Southeast Asian Flying Snake	assessed
55	***Lycodon aulicus* (Linnaeus, 1758)**	Common Wolf Snake	Non-venomous Common, Not assessed
56	***Lycodon travancoricus* (Beddome, 1870)**	Travancore Wolf Snake	Non-venomous Common, Least Concern
57	***Dryocalamus nympha* (Daudin, 1803)**	Common Bridal Snake	Non-venomous Rare, Not assessed
58	***Dryocalamus gracilis* (Gunther,1864)**	Gunther's Bridal Snake, Slender Bridal Snake, Scarce Bridal Snake	Non-venomous Data deficient
59	***Sibynophis Subpunctatus* (Dumeril, Bibron & Dumeril, 1854)**	Dumeril's Black-headed Snake	Non-venomous Uncommon, Not assessed
60	***Xenochrophis piscator* (Schneider, 1799)**	Checkered Keel-back, Common Water Snake	Non-venomous Common, Not assessed
61	***Amphiesma stolatum* (Linnaeus, 1758)**	Striped Keelback Snake	Non-venomous Common, Not assessed
62	***Amphiesma beddomei* (Gunther, 1864)**	Beddome's Keelback, Nilgiri Keelback	Non-venomous Common, Least Concern
63	***Amphiesma monticola* (Jerdon, 1853)**	Yellow-collared Forest Keelback, Hill Keelback	Non-venomous Uncommon, Least Concern
64	***Macropisthodon plumbicolor* (Cantor, 1839)** ***Macropisthodon plumbicolor plumbicolor* (Cantor, 1839)**	Green Keelback, Indian Green Keelback	Non-venomous Uncommon, Not assessed
65	***Atretium schistosum* (Daudin, 1803)**	Olivaceous Keelback, Olive Keelback Water Snake, Split Keelback Snake	Non-venomous Uncommon, Not assessed
66	***Rhabdops olivaceus* (Beddome, 1863)**	Olive Forest Snake, Olive Trapezoid Snake	Non-venomous Rare, Least Concern
67	***Xylophis perroteti* (Dumeril, Bibron & Dumeril, 1854)**	Striped Small-headed Snake, Striped Narrow-headed Snake, Perrotet's Mountain Snake	Non-venomous Uncommon, Least Concern

Table 1: contd…

68	***Xylophis captaini* (Gower & Winkler, 2007)**	Captain's Wood Snake	Non-venomous, Common, Least Concern
69	***Boiga trigonata* (Bechstein, 1802)** ***Boiga trigonata trigonata* (Bechstein, 1802)**	Common Cat Snake, Indian Gamma Snake	Mildly venomous Common, Least Concern
70	***Boiga forsteni* (Dumeril, Bibron & Dumeril, 1854)**	Forsten's Cat Snake, Reddish Peninsular Cat Snake	Mildly venomous Common, Least Concern
71	***Boiga ceylonensis* (Gunther, 1858)**	Ceylon Cat Snake	Mildly venomous Common, Not assessed
72	***Boiga dightoni* (Boulenger, 1894)**	Travancore Cat Snake, Peerimade Cat Snake, Dighton's Cat Snake	Mildly venomous Data deficient
73	***Ahaetulla nasuta* (Bonnaterre, 1790)**	Common Green Whip Snake, Common Vine Snake	Mildly venomous Common, Not assessed
74	***Ahaetulla dispar* (Gunther, 1864)**	Gunther's Whip Snake, Gunther's Vine Snake	Mildly venomous Common, **Near Threatened**
75	***Ahaetulla perroteti* (Dumeril, Bibron & Dumeril, 1854)**	Bronze-headed Vine Snake, Perrotet's Vine Snake, Western Ghats Bronzeback	Mildly venomous Rare, **Endangered**
76	***Ahaetulla pulverulenta* (Dumeril, Bibron & Dumeril, 1854)**	Brown Whip Snake, Brown Vine Snake, Brown-speckled Whipsnake, Thunderbolt Snake	Mildly venomous Common, Least Concern
77	***Enhydris sieboldii* (Schlegel, 1837)**	Siebold's Smooth-scaled Water Snake	Mildly venomous Common, Least Concern
78	***Dieurostus dussumierii* (Dumeril, Bibron & Dumeril, 1854)**	Dussumier's Smooth Water Snake	Mildly venomous Common, Least Concern
79	***Cerberus rhynchops* (Schneider, 1799)**	Dog-faced Water Snake	Mildly venomous Common, Least Concern
80	***Gerardia prevostiana* (Eydoux & Gervais, 1837)**	Glossy Marsh Snake	Mildly venomous Uncommon, Not assessed
Family: Elapidae Boie, 1827			

Table 1: contd…

81	***Bungarus caeruleus* (Schneider, 1801)**	Common Krait	Highly venomous Common, Not assessed
82	***Calliophis melanurus* (Shaw, 1802)**	Slender Coral Snake, Indian Slender Coral Snake	Highly venomous Rare, Not assessed
83	***Calliophis nigrescens* (Gunther, 1862)**	Striped Coral Snake, Black Coral Snake	Highly venomous Rare, Least Concern
84	***Calliophis bibroni* (Jan, 1858)**	Bibron's Coral Snake, Bibron's Slender Coral Snake	Highly venomous Rare, Least Concern
85	***Naja naja* (Linnaeus, 1758)**	Common Cobra, Indian Cobra, Binocellate, Spectacled Cobra	Highly venomous Common, Least Concern
86	***Ophiophagus hannah* (Cantor, 1836)**	King Cobra, Hamadryad	Highly venomous Rare, **Vulnerable**
Family: Hydrophiidae Boie, 1827			
87	***Hydrophis fasciatus* (Schneider, 1799)** ***Hydrophis fasciatus fasciatus* (Schneider, 1799)**	Banded Sea Snake, Striped Sea Snake	Highly venomous Common, Least Concern
88	***Hydrophis spiralis* (Shaw, 1802)**	Yellow Sea Snake	Highly venomous Common, Least Concern
89	***Hydrophis ornatus* (Gray, 1842)** ***Hydrophis ornatus ornatus* (Gray, 1842)**	Cochin Banded Sea Snake, Ornate Sea Snake	Highly venomous, Rare, Least Concern
90	***Hydrophis cyanocinctus* Daudin, 1803**	Annulated Sea Snake, Blue Banded Sea Snake	Highly venomous Common, Least Concern
91	***Hydrophis gracilis* (Shaw, 1802)**	Common Narrow-headed Sea Snake,Common Small-headed Sea Snake, Graceful Small Headed Sea Snake, Slender Sea Snake	Highly venomous Common, Least Concern
92	***Hydrophis cantoris* Gunther, 1864**	Cantor's Narrow-headed Sea Snake, Gunther's Sea Snake	Highly venomous Data deficient
93	***Enhydrina schistosa* (Daudin, 1803)**	Hook-nosed Sea Snake, Beaked Sea Snake	Highly venomous Common, Least Concern
94	***Lapemis curtus* (Shaw, 1802)**	Malabar Sea Snake, Short Sea Snake, Spine Bellied Sea Snake	Highly venomous Common, Least Concern

Table 1: contd…

95	***Kerilia jerdonii* (Gray, 1849)** ***Kerilia jerdonii jerdonii* (Gray, 1849)**	Jerdon's Sea Snake	Highly venomous Uncommon, Not assessed
96	***Praescutata viperina* (Schmidt, 1852)** ***Thalassophina viperina* (Schmidt, 1852)**	Viperine Sea Snake	Highly venomous Common, Least Concern
97	***Astrotia stokesii* (Gray, 1846)**	Large headed Sea Snake, Stoke's Sea Snake	Highly venomous Common, Least Concern
Family: Viperidae Boie, 1827			
98	***Daboia russelii* (Shaw & Nodder, 1797)** ***Daboia russelii russelii* (Shaw & Nodder, 1797)**	Russell's Viper	Highly venomous Common, Least Concern
99	***Echis carinatus* (Schneider, 1801)** ***Echis carinatus carinarus* (Schneider, 1801)**	Saw-scaled Viper	Highly venomous Common, Not assessed
100	***Hypnale hypnale* (Merrem, 1820)**	Hump-nosed Pit Viper	Highly venomous Common, Not assessed
101	***Trimeresurus malabaricus* (Jerdon, 1854)**	Malabar Rock Pit Viper, Malabar Pit Viper	Highly venomous Common, Least Concern
102	***Trimeresurus gramineus* (Shaw, 1802)**	Green Pit Viper, Bamboo Pit Viper, Common Bamboo Pit Viper	Highly venomous Common, Least Concern
103	***Trimeresurus macrolepis* (Beddome, 1862)** ***Peltopelor macrolepis* (Beddome, 1862)**	Large scaled Pit Viper	Highly venomous Common, **Near Threatened**
104	***Trimeresurus strigatus* Gray, 1842**	Horse-shoe Pit Viper	Highly venomous Common, Least Concern

* based on IUCN Red List of Threatened species 2013.1 version

Figure 2.1: *Ramphotyphlops braminus.*

Figure 2.2: *Grypotyphlops acutus.*

Figure 2.3: *Uropeltis ceylanica.*

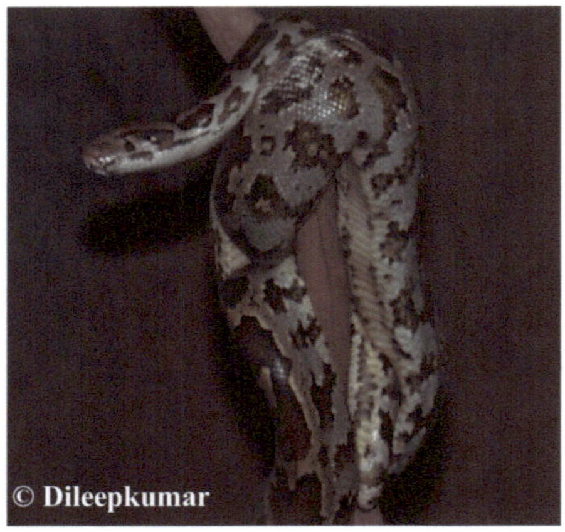

Figure 2.4: *Python molurus.*

Figure 2.5: *Gongylophis conicus.*

Figure 2.6: *Acrochordus granulatus.*

Figure 2.7: *Coelognathus helena.*

Figure 2.8: *Ptyas mucosa.*

Figure 2.9: *Oligodon arnensis.*

Figure 2.10: *Dendrelaphis tristis.*

Figure 2.11: *Chrysopelea ornate.*

Figure 2.12: *Lycodon aulicus.*

Figure 2.13: *Xenochrophis piscator.*

Figure 2.14: *Amphiesma stolatum.*

Figure 2.15: *Xylophis captaini.*

Figure 2.16: *Ahaetulla nasuta.*

Figure 2.17: *Dieurostus dussumierii.*

Figure 2.18: *Cerberus rhynchops.*

Figure 2.19: *Bungarus caeruleus.*

Figure 2.20: *Calliophis melanurus.*

Figure 2.21: *Calliophis bibroni.*

Figure 2.22: *Naja naja.*

Figure 2.23: *Ophiophagus Hannah.*

Figure 2.24: *Hydrophis ornatus.*

Figure 2.25: *Daboia russelii.*

Figure 2.26: *Echis carinatus.*

Figure 2.27: *Hypnale hypnale.*

Figure 2.28: *Trimeresurus malabaricus.*

Figure 2.29: *Trimeresurus macrolepis.*

3. THREATS TO DIVERSITY OF SNAKES

Snakes are ecologically significant reptilian groups both as predator and prey, but very little measures have been taken towards exploring their diversity. The current status of these reptilian taxa is that they are declining all around the world [1-4]. Even though there an abundance in species diversity in Western Ghats region, a critical turn down in the population of many snake species has been encountered and it is necessary to tackle this threat of decline in their population. The narrow distribution ranges of snakes [5, 6] are making them more susceptible. For the development of targeted conservation guidelines, it is obligatory to understand the current ecological status of individual snake species along with its ecological niche.

Snakes and the other reptilian groups in the rain forest of South Western Ghats region have declined significantly which may be due to the destruction, degradation, fragmentation and the loss of habitat by anthropogenic disturbances like wild life tourism, construction of roads, clearing of forests (Fig. **3**), fire, plantations *etc.* The intact habitats in the forests that were once unscathed by humans have changed radically in the recent decades primarily for the human settlemental needs (Fig. **4**) and for the expansion of industrialized areas. This resulted in the sequential vanishing of many floral and faunal species from the site.

Figure 3: Logging of tropical rainforests continues to be a major source of habitat destruction.

In addition to the lack of studies on snake faunal diversity in this biome, scarcities of research are evident in conservation planning and management field also. Lack of data particularly on population size and species-specific habitat needs are the chief limitations facing conservation strategies [7]. The fact that Western Ghats region is a hotspot with limited studies strongly suggest the need for systematic taxonomical studies of snake populations and its habitat relationships, thereby enhancing the understanding of snake ecology and required conservation measures. The rapid rate of degradation of natural ecosystem leads to habitat loss of different species of snakes. The high risk of persecution of many of these species may be underestimated due to the deficiency of exact population information. Among the known species in this hotspot region many are included in data deficient category by IUCN and some are in threatened and endangered Red list.

Figure 4: Land use change of rainforest for rubber plantation.

Global statistics on current threat status of reptiles indicates that one in five species are threatened with extinction, while another one in five species are data deficient. [8]. The King Cobra is threatened due to habitat loss and over-exploitation for trade. IUCN reports a 30% of reduction in King cobra population from India over 75 years due to numerous threats including habitat destruction and harvesting of mature individuals from the wild [9]. A study report published by the Reptilian Research Group, University of Kerala has revealed the highest threat status for King cobra inhabiting the Western Ghats. The study indicates that the migration of King cobra from their natural habitat to nearby villages may be due to large-scale deforestation and the disturbances caused by poachers [10]. Another threat to conservation management is the limited distribution range of rare snake species which makes population estimation difficult. Dubious loss and fragmentation of habitat has reduced its numbers. Unpredicted forest fires (Fig. **5**) pose another threat to terrestrial snake species in the wild.

Figure 5: Forest fires occur in the rainforests, a threat to the natural habitat.

The quality loss of habitat is not only observed in terrestrial ecosystems, but also in aquatic ecosystem due to anthropogenic activities, ill-effects of global warming and climate changes in the recent decades. This forced the aquatic reptiles to migrate from their natural habitats to other areas, where they weren't normally known to reside in. Snakes as a whole have very little commercial value as compared with other group of animals, but they are exploited in various ways. Even though snakes are listed under 'Scheduled category' by Wild life protection law in India, snakes are illegally harvested for venom, medicinal purposes, food and also for skin trade. In many parts of the world snakes suffer greatly at the hands of motorists; their elongated bodies and relatively slow movement makes them easy targets, whether intentional or accidental, In addition, many linger on the warm tarmac surfaces during their nocturnal wanderings. Clearing of forests raises the ground temperature and dries the land which affects the snakes adversely, especially the burrowers.

Local reports indicated that 10-15 snakes are captured daily by snake catchers from human settlement areas in Kerala. This data strongly implies snake migration due to natural habitat loss. Migration poses many threats to the species residing in the restricted habitat such as food availability, proper space for rearing their eggs, mating space, territory maintenance *etc.* In addition, road accidents (Fig. **6**), intentional human attacks due to fear are further threats faced by these migratory snakes. Snakes have to compromise with their cryptic behaviour like hibernation, feeding and egg-laying. The physical structure and thermal properties of the inhabiting sites are crucial to these endothermic creatures than the floristic composition of habitat; hence the continuous decline in the quality of habitat is the fundamental cause that makes species vulnerable.

Figure 6: A road accident leads to the death of *Uropeltis* sp.

The scientific community must take responsibility for the large scale exploitation of snakes in the past. Museums and universities throughout the world are often crammed with preserved specimens, including many that have never been properly studied or even classified (Fig. **7**).

Taxonomists and anatomists failed to cooperate with one another and so many collections were duplicated unnecessarily. Even recently published papers have reported the killing of large numbers of individuals for the sake of research, sometimes to confirm facts that are already well known.

Figure 7: A common scene of museums with display of formalin/alcohol fixed specimens.

4. EFFECTIVE CONSERVATION STRATEGIES FOR SNAKES

From conservation point of view, species specific approach is useful for the effective conservation planning and management. The species specific approach is effective in managing the population in a region because each species is found to be the vital part of the ecosystem either as a prey or as a predator. Variation in species-habitat relationship is an essential component of conservation and has been emphasized at different scales since the late 1980s [11-13]. Scrupulous research on the diversity and the population density of snake species along with thorough understanding on the ecological status is essential [14, 15]. Recent advances in computational methods have improved multiscale examination of species-habitat associations [15, 16] Various studies suggest differences in environmental factors that influence species distribution and species abundances [17]. In order to highlight urgent conservation measures for snakes, it is needed to identify the critical snake species and habitats for conservation. Conservation area networks (CANs) can be set up considering species distributions and also with the help of using algorithms that aim to maximize biodiversity representation in a specific area [18].

Agricultural development has resulted in land use changes [19-21]. The snake distribution along the urban-agricultural gradient varies and this suggests adopting different management strategies based on surrounding land cover. A case study on snake species distributions from the American tallgrass prairie recommended the use of a species specific approach to manage snake populations [22]. In most places biodiversity conservation goes along with conservation of agricultural landscapes [23], because some species depend upon farming lands to meet their habitat requirements, as they have coevolved with agriculture [24].

The modifications in the usage of agricultural farm land over the last 50 years has resulted in spread of secondary vegetation with negative effects on biodiversity [25, 26]. The conversion of natural ecosystems to croplands is likely to be most threatening to global biodiversity [27].

Hence, the primary challenge for the conservation biologists is to develop approaches for the establishment of secluded areas that satisfactorily protect biodiversity [28]. As the population status of most species is known without firm data on population genetics, large scale field experiments and continuous monitoring to assess the population trends within the specific site are urgently needed in order to understand the distribution pattern and the ecological status of different snake species. Snakes surveys including visual searches and refuge survey are the critical steps in the habitat management planning. Reptiles are particularly vulnerable to declining habitat quality and inappropriate habitat management. The habitat management requirements of invertebrates and reptiles are very similar, providing a wide range of microhabitat within a site. Zoogeographic and geomorphic factors are critically influencing factors for species richness and community composition. A mid-successional habitat is generally preferable for the reptilian groups and snakes require an open area for warmth and more vegetated areas for shelter. Outside the protected areas, conservation can also extend to sites like farm lands, churchyards, utility sites in the local areas *etc*. Captive breeding programmes may be useful where certain species' habitats are seriously degraded or have disappeared entirely. There should be an ultimate goal of restoring the species to its natural habitat with genetic integrity.

All Indian snakes are protected under the Wildlife (Protection) Act, 1972 and none can be collected, kept, transported or exported without the relevant permits from the Forest Department. The Wildlife Act had several orders of protection, Schedule I to IV, and the offences involving these Scheduled snakes are punishable with a fine. So strictly prohibiting commercial collection of snakes may help to stem the flood of certain species that reach the pet trade in huge numbers. Collection of skin is another area that could be similarly controlled. Regulations can in a way restrict the snake trade both domestic and internationally.

Perhaps the best means of conservation is education. An urban population is unlikely to seriously consider setting aside areas of land for snakes while current prejudices prevail. Zoos and snake parks have an important role to play and there has been a noticeable change in attitudes. Amateur herpetological societies by putting on responsible demonstrations and displays to the public can encourage young members. Undoubtedly, snake conservation and habitat management deserves more attention and conservation protocols needs to be refined in future.

ACKNOWLEDGEMENTS

We are deeply grateful to the Principal Chief Conservator of Forest & Wildlife of Kerala for granting permission to our snake research activities and State Inter University Centre for Excellence in Bioinformatics (SIUCEB) for research funding. We would particularly like to express our heartfelt thanks to Dr. Achuthsankar S. Nair, Head of the Department, Department of Computational Biology and Bioinformatics, University of Kerala for the facilities provided to complete the work.

CONFLICT OF INTEREST

The authors confirm that this chapter contents have no conflict of interest.

REFERENCES

[1] Mount RH. The Reptiles and Amphibians of Alabama. Auburn University Agricultural Experiment Station, Auburn, Alabama 1975.

[2] Dodd Jr CK. Status, conservation, and management. In: Seigel, RA *et al,* Eds. Snakes: Ecology and Evolutionary Biology. MacMillan, New York 1987.

[3] Gibbons JW, Scott DE, Ryan JT, Buhlmann KA, Tuberville TD, Metts B, Greene JL, Mills TM, Lediden YA, Poppy SM, Winne CT. The global decline of reptiles: dejavu amphibians. BioScience 2000; 50: 653-666.

[4] Filippi E, Luiselli L. Changes in community composition, habitats and abundance of snakes over 10+ years in a protected area in Italy: conservation implications. The Herpetological Journal 2006; 16: 29-36.

[5] Anderson S. Aerography of North American fishes, amphibians, and reptiles. American Museum Novitates 1984; 28: 02, 1-6.

[6] Anderson S, Marcus LF. Aerography of Australian Tetrapods. Australian Journal of Zoology 1992; 40: 627-651.

[7] Dodd Jr CK. Strategies for snake conservation. In: Seigel RA, Collins JT, Eds. Snakes: Ecology and Behavior. Mc-Graw-Hill, New York 1993.

[8] Böhm Monika, Reynolds, Robert P *et al.* The conservation status of the world's reptiles Biological Conservation 2013; 157: 372-385.

[9] Stuart B, Wogan G, Grismer L, Auliya M, Inger RF, Lilley R, Chan-Ard T, Thy N, Nguyen TQ, Srinivasulu C, Jelić D. Ophiophagus hannah. The IUCN Red List of Threatened Species. Version 2014.2 2012.

[10] Dileepkumar R, Suresh B, Baiju S, Anaswara Krishnan S, Divya L, Oommen OV. Habitat loss forcing king cobras (*Ophiophagus hannah*) to migrate to human habitats? Ind. Biodiv. Cong. Abst 2010; 1: 67-68.

[11] Morris DW. Ecological scale and habitat use. Ecology 1987; 68: 362-369.

[12] Wiens JA, Rotenberry JT, Van Horne B. Habitat occupancy patterns of North America shrubsteppe birds: the effects of spatial scale. Oikos 1987; 48: 132-147.

[13] Wiens JA. Spatial scaling in ecology. Functional Ecology 1989; 3: 358-397.

[14] Rushton SP, Ormerod SJ, Kerby G. New paradigms for modelling species distributions? Journal of Applied Ecology 2004; 41: 193-200.

[15] Guisan A, Thuiller W. Predicting species distribution: offering more than simple habitat models. Ecology Letters 2005; 8: 993-1009.

[16] Guisan A, Zimmerman NE. Predictive habitat distribution models in ecology. Ecological Modelling 2000; 135: 147-186.

[17] Nielsen SE, Johnson CJ, Heard DC, Boyce MS. Can models of presence-absence be used to scale abundance? Two case studies considering extremes in life history. Ecography 2005; 29: 197-208.

[18] Sarkar S, Illoldi-Rangel P. Systematic conservation planning: an updated protocol. Natureza Conservação 2010; 8: 19-26.

[19] Hobbs RJ, Saunders DA. Effects of landscape fragmentation in agricultural areas. In: Moritz CC, Kikkawa J. Eds. Conservation Biology in Australia and Oceania. Surrey Beatty and Sons, Chipping Norton, NSW 1994.

[20] Fahrig L. Effects of habitat fragmentation on biodiversity. Annu. Rev. Ecol. Evol. Syst. 2003; 34: 487-515.

[21] Lindenmayer DB, Fischer J. Habitat Fragmentation and Landscape Change: An Ecological and Conservation Synthesis. Island Press, Washington, DC 2006.

[22] Nicolette L Cagle. Snake species distributions and temperate grasslands: A case study from the American tallgrass prairie Biological Conservation 2008; 141: 744-755.

[23] Henle K, Alard D, Clitherowc J, Cobb P, Firbank L, Kull T, McCracken D, Moritz RFA, Niemela J, Rebane M, Wascher D, Watt A, Young J. Identifying and managing the conflicts between agriculture and biodiversity conservation in Europe-a review. Agric. Ecosyst. Environ 2008; 124: 60-71.

[24] Atauri JA, de Lucio JV. The role of landscape structure in species richness distribution of birds, amphibians, reptiles and lepidopterans in Mediterranean landscapes. Land. Ecol 2001; 16: 147-159.

[25] Ramankutty N, Foley JA. Estimating historical changes in global land-cover: croplands from 1700-1992. Global Biogeochem. Cy 1999; 13: 997-1027.
[26] Kanowski JJ, Reis TM, Catterall CP, Piper SD. Factors affecting the use of reforested sites by reptiles in cleared rainforest landscapes in tropical and subtropical Australia. Restor. Ecol 2006; 14: 67-76.
[27] Sala OE. Global Biodiversity Scenarios for the Year 2100. Science 2000; 287: 1770-1774.
[28] Margules CR, Pressey RL. Systematic conservation planning. Nature 2000; 405: 243-253.

CHAPTER 2

Conservation of Medicinal Plant Resources through Community Born Biodiversity Management Committee, Kerala, India

Laladhas K. P.[1,2,*], Preetha N.[1], Baijulal B.[1] and Oommen V. Oommen[1]

[1]*Kerala State Biodiversity Board, L-14, Jai Nagar, Medical College P.O, Thiruvananthapuram-695 011, Kerala, India and* [2]*Life Science Research Centre, St. Stephans College, Pathanapuram, Kollam, Kerala, India*

Abstract: Sustainable harvest of Medicinal and Aromatic species should become an important conservation strategy for the long term survivability of medicinal plant resources. Presently medicinal plants are mainly collected from the wild (90%) and 70% is through destructive harvesting. This paper presents a community model policy involving Medicinal and Aromatic Plants (MAP) Resource management by Local Self Government (LSG) & Biodiversity Management Committee (BMC), which will ensure sustainable utilization of medicinal plants and fair benefit sharing. Stricter enforcement of biodiversity legislations and mandatory registration for collectors from wild and herbal shops is recommended with proper inventory, and monitoring of harvesting impacts. Focus is on strengthening LSG & BMC for Resource Management and formation of medicinal plants trading society. The collected materials are to be properly tagged with local and trade name, and source of collection in order to support traceability to collection area. It is recommended that Good Manufacturing Practices (GMP) certification should be linked with traceability of MAP and the development of Live Green label for materials harvested sustainably is proposed. Sustainable harvesting of MAP is intricately connected with conservation of unique habitats benefitting the local ecosystem greatly and facilitates development of a strong sense of stewardship in the communities which would help in sustainable management of the habitats and biodiversity conservation at the same time ensuring rural livelihood.

Keywords: Ayurvedic industry, benefit sharing, biodiversity legislations, biodiversity registers, commercial utilization, conservation, destructive harvesting, fair trade, GMP certification, habitat loss, herbal shops, medicinal plants, over-exploitation, quality medicines, resource management, scientific management, sustainable harvest, trading society, traditional knowledge, traditional remedies.

***Corresponding author Laladhas K.P.:** Kerala State Biodiversity Board, L-14, Jai Nagar, Medical College P.O, Thiruvananthapuram- 695 011, Kerala, India; Tel: 0471 - 2559134 (off.); Fax: 0471 – 2448234; Mob: 9447271034; E-mail: laladhas@gmail.com

Laladhas K.P., Oommen V. Oommen and Sudhakaran P.R. (Eds.)

1. INTRODUCTION

The "World Health Organization estimates that over 80 per cent of the world's population depends on traditional plant-based medicine for their primary health care concerns" [1]. The global demand for natural health care products is increasing due to their non-toxic nature and negligible side effects. The domestic market for herbal drugs "are growing at an annual rate of 20 per cent while the international market for medicinal plant-based products is estimated to be growing at 7 per cent per annum" [2]. "Globally, wild or natural resources meets 70 to 90 percent of the market demand for Medicinal and Aromatic Plants (MAP)" [3]. Commercial harvesting has threatened the availability of more than 15,000 medicinal plants globally [4]. The loss of biodiversity of MAP means loss of new potential drugs as also the loss of traditional medicines for meeting the health concerns of local communities. MAP is also related to health, income and cultural identity of the rural society.

China and India are two great producers of medicinal plant based products [5]. India is one of the 17 mega biodiverse areas of the world [6]. India has areas with all three biomes hills, plateaus and coasts representing different habitat diversity making it one of the richest sources of plant biodiversity.

The domestic market for ayurvedic, products is about Rs. 3000 crores, which is growing at a rate of 15-20% p.a. out of which 80% is accounted for by Over the Counter (OTC) products [7]. "India, at present, exports herbal material and medicines to the tune of Rs. 1210 crores annually, comprising of Rs. 593 crores from saps and extracts, Rs. 370 crores from plants and plant parts, and Rs. 235 crores from Ayurvedic and Unani medicines" [7]. "There are about 6000 licensed pharmacies of Indian System of Medicine in addition to Homeopathy and a number of unlicensed small scale processing units engaged in the manufacture of medicines" [2]. About 881 plant species are used in herbal products, of which 60 species are imported, 60 species are cultivated and 760 species are collected from the wild. Of the approx. 500 medicinal plant species used by the ayurvedic industry, around 80 per cent are procured from wild areas" [8]. "Due to habitat loss and over-exploitation, approximately 1,000 medicinal species are under threat in India" [4]. Presently medicinal plants are mainly collected from the wild (90%) and 70% is through destructive harvesting [9]. Kerala has been one of the leaders of ayurvedic drugs in India.

2. GOOD COLLECTION PRACTISES

It is mandatory to develop a sustainable and good collection practice for MAP harvested from the wild. The stress is on sustainable practices and includes industries and shops selling raw medicinal plants/parts and industries. Sustainable harvesting will ensure conservation of genetic stock, and uniform future availability, improved quality of MAP and conservation of biodiversity. The model is formulated keeping in mind that the collectors are sections of people who have good knowledge of MAP resources, its availability, quality and quantity at wild. The model is also formulated with the objective of improved benefit sharing and creating awareness to conserve, manage and collect MAP's responsibly.

The broad objectives are:

1. To outline principles and criteria for management and monitoring of harvesting of MAP species.
2. Review of Current status of medicinal plants industry, availability, practices adopted by collectors, trends in trade and existing regulations.
3. Strategies for increasing awareness about sustainable management practises.
4. Utilization of baseline data of MAP diversity from People's Biodiversity Register (PBR) and mapping/survey of collection sites.
5. Enlistment for Rare endangered and threatened (RET) varieties, diversity and density.
6. Strategy for conservation and sustainable use of medicinal plants in its habitat.
7. Guidelines for good collection practices especially for those species for which destructive collection is employed.
8. Guidelines for Good sourcing practices for industries.
9. Capacity development at local level.

10. Local level benefit sharing to the people who really conserve MAP.

3. DISTRIBUTION OF MEDICINAL PLANTS

"Around 70% of India's medicinal plants are found in tropical areas mostly in the Western and Eastern Ghats, the Vindhyas, Chotta Nagpur plateau, Aravalis & Himalayas" [5]. "Majority of the medicinal plants are higher flowering plants, around 24 per cent of species occur as trees, 20 per cent as shrubs and 56 per cent as herbs" [8].

4. CONSTRAINTS IN THE INDUSTRY

The major causes of loss of biodiversity are habitat destruction, pollution, climate change, change in agricultural land use, changes in water resources, introduction of non native invasive species, and over-exploitation of medicinal plants and lack of monitoring system. Traditional herbal medicines were produced using knowledge handed over through generations by practitioners. The present day system of indiscriminate collection of MAPs from wild sources has depleted these valuable resources. "A species which (i) has a narrow geographic distribution, (ii) is habitat- specific, and (iii) has small population sizes everywhere, is more easily over-harvested than other species" [10]. Roots accounts for 30% of trade and the threat of overexploitation of the species depends specifically on the plant part of medicinal value. "70% of the plant collections involve destructive harvesting by the use of parts like roots, bark, wood, stem and the whole plant in case of herbs" [5]. The natural regeneration capacity of the species is affected leading to a dwindling of the genetic stock.

Post harvesting, the plant material is stored poorly and 40% of the material is wasted due to spoilage during storage and transportation. Lack of established grading norms, leads to non- uniform quality of plant material and effects market prospects. As the price paid to the collectors by the middleman tend to be lower than the market price overexploitation of the natural resources much beyond the natural regeneration capacity of the species occurs.

5. GOOD QUALITY MAP RESOURCES NEEDED FOR QUALITY MEDICINES

The quality of medicinal plant parts is based on the geographical origin, time of collection, maturity and post harvest handling. Plants are collected without much awareness of optimum stage of maturity, degree of permissible moisture content or the proper storage condition for maintaining the shelf life of the active ingredients. "Collectors pay little attention to quality and continue to collect prematurely" [11]. Market information and documentation lack reliability because traders are not legally required to provide details of collection, source, and methods of drying and processing of wild collected materials.

6. WILD COLLECTION *IN SITU* COMMERCIAL CULTIVATION

The practice of collection of MAP from wild is expected to continue due to several reasons. The plants are harvested largely from wild populations [12]. Large scale cultivation of medicinal plants is not possible mainly because of lack of sufficient knowledge about the propagation aspects and non availability of planting materials of most of the species of commercial interest. The absence of sufficiently reliable market and relatively low market volume for some species and uncertainty in trade hampers commercial cultivation.

"Cultivated plants are sometimes qualitatively inferior when compared with wild gathered specimens. Thus wild ginseng roots are 5-10 times more valuable than roots produced by artificial propagation" [13]. "The cultivation of MAP species reduces the economic incentives for forest-dependent people to conserve the ecosystems in which the MAP species occur" [14]. Sustainable harvesting of wild harvested species and their habitat is assuming greater significance in this context.

7. CHARACTERISTICS OF MARKET

Generally the raw material collected from the wild by primary collector is passed through middleman who bid for the right to collect from specified areas. Medicinal plants collected from the wild by tribals, and other local people are marketed by the traders in towns and ultimately are brought by drug manufacturers. Medicinal plants trade is secretive, informal, largely unorganized and unregulated. This supply chain often extends to 3-4 tiers without much value addition but with increase in sale price at each level. An increase of 70-300 % of sale price occurs from primary collector's level to end used.

In Kerala, the marketing of medicinal plants is undertaken by the Kerala State Federation of Scheduled Caste and Scheduled Tribes Development Co-operatives Limited.

8. EXISTING GUIDELINES AND POLICIES IN INDIA - THE BIOLOGICAL DIVERSITY ACT, 2002, AND RULES, 2004

The Indian System of Medicine has adopted the Good Manufacturing Practices (GMP) for herbal products manufacturing units, for quality assurance. Fair trade certification in India exists for various plant produce and respects the rights of local communities and recognizes traditional knowledge. But it has not been widely accepted for medicinal plants as this sector is mostly unorganized. Schedule T of Drugs and Cosmetics Act, 1940 specifies that the raw materials should be of good quality for GMP certification.

As per the Indian 'The Biological Diversity Act, 2002, and Rules, 2004' (Section 7) prior intimation to State Biodiversity Board for obtaining biological resources for commercial utilization is necessary for corporate, associations or organizations registered in India. Commercial utilization means use of biological resources for drugs, industrial enzymes, food flavours, cosmetics, oleoresins, colours. But this does not apply to the local people or communities.

8.1. BMC- Roles and Responsibilities

Section 41(1) of 'The Biological Diversity Act, 2002, and Rules, 2004' "stipulates that every local body shall constitute a BMC for promoting conservation, sustainable use and documentation of biological diversity". The National Biodiversity Authority (NBA) and State Biodiversity Boards shall consult the BMC while taking any decision relating to use of biological resources. In section 41(3) it is also stipulated that BMC "may levy charges by way of collection fees from any person for collecting biological resources for commercial purposes for areas within its jurisdiction". But due to the unaccounted nature of such resources the stakeholders are not in a position to comply with this.

9. NECESSITY FOR AN ACTION PLAN

A scientific management system for harvesting from the wild is needed. The management system must include annual harvest quotas; seasonal or geographical restrictions and restriction of harvest to particular plant parts or size. "Harvesting techniques need to be formulated as the extraction of the roots or bark is often negatively affecting the regeneration of the species. Collection methods are often crude and wasteful, resulting in loss of quality and reduction in price" [15]. In majority of cases the collector or the contractor who extract medicinal plant products from the wild does not have responsibility for sustainable harvesting. Drug manufacturers should be encouraged to use certified raw materials. Legal collection, sustainable resource management, good quality and traceability of raw materials should be the social responsibility of corporates. Developing standards and complying with them should be a collective responsibility.

10. ACTION PLAN FOR CONSERVATION, SUSTAINABLE AND EQUITABLE SHARING OF WILD MAP RESOURCES AT LSG LEVEL IN INDIA - A MODEL APPROACH

The paper aims to promote sustainable management of collection of MAP from the wild for commercial purposes by strengthening Local Self Government (LSG) & BMC for monitoring. The Action plan specifies formation of a Medicinal Plants trading society, an approved society under a committee comprising of LSG and BMC. The society will buy the collected materials at specified price and shops and pharmaceuticals can buy from them. The BMC will plan and monitor the management of harvesting of MAP by registered collectors (Fig. **1**). The organizational structure includes 1) Multi disciplinary team comprising of Resource management agency (Kerala State Biodiversity Board, KSBB), 2) Resource facilitating institutions (Medicinal Plants Board, Forest department *etc.*) for policy decisions 3) Resource management by members of LSG & BMC 4) A MAP trading society. Fig. **2**. depicts the schematic representation of proposed supply chain of MAP resources. Table **1**. illustrates the roles and responsibilities of resource management agencies and Table **2**. gives the details of action plan components.

Table 1: Roles and responsibilities of resource management agencies

Sl. No	Resource Management Agencies	Responsibilities
1	Resource Management Agency (KSBB)	Policy matters, Technical knowhow, Market trends
2	Resource Facilitating institutions (Medicinal Plants Board, Forest dept)	Technical Knowhow, Co-ordination
3	Resource management by LSG & BMC	Harvesting Permits, Formulate Scientific & Good Collection practices, Registration of collectors, Monitoring and harvesting
4	Collectors	Registered Collectors, Good collection practices
5	Shops	Compliance with rules, Fair trade, Proper storage and traceability practices
6	Industries	Good sourcing practices, Benefit sharing, Live Green Label on products

Table 2: Components of Action Plan

Sl.No	Principles	Method of Implementation
1.	Conservation of RET	1. Rare, endangered and threatened plants in the local area to be identified using PBR 2. No permits for Endangered or critically endangered species. 3. Site-specific strategies should be worked out for protection of such species
2.	Sustainable collection of MAP from wild	1. A LSG level species wise/habitat wise management plan based on good collection practices. Participatory approach involving local community in management.
	2.1 Harvesting	1. BMC may be entrusted for managing the collection of resources 2. BMC to monitor and evaluate the management of harvesting by Providing harvesting permits to collectors 3. Harvesting permit can specify the species, plant parts. 4. Harvesting permit to specify amount, harvesting technique
	2.2 User rights	1. Collection site within the boundary of LSG where collection is to be carried out to be specified in harvesting permit.
	2.3 Minimizing wastage	BMC to ensure that wastage of MAP resources due to wrong collection practices is minimized
	2.4 Monitoring	BMC to monitor the population of harvested, non harvested, species impact of harvesting and to ensure that regeneration rate exceeds the collection rate
3.	Proper Collection practices	Good collection Practices to be formulated by BMC based on local conditions especially for those for which destructive collection is being practiced
	3.1Scientific calendar	1. The population density of the target species being collected commercially from the LSG should be determined and a collection plan that maintains the base population is to be adopted. 2. The optimum season for collection should be determined according to the quality and quantity of biologically active constituents as also total vegetative yield of the targeted medicinal plant parts. 3. A scientific calendar specifying methods as frequency, Maximum allowed collection intensity for species to be made available to collectors by BMC. This can be formulated based on the propagation cycle of the species as per annual growth rate. 4. Minimum and maximum maturity and size of collected parts allowed for the target species to be defined in instructions.
	3.2 Non destructive method of collection	1. Depending on the species and plant part collected a non-destructive harvesting should be practised 2. When collecting roots of trees and bushes, damage to the main root should be avoided. Similarly for other plant parts appropriate standards to be established.
	3.3 Analogues and adulterants	1. Maximum care should be taken to avoid adulteration or contamination from foreign materials. 2. The scientific calendar shall also specify analogous plants available locally which may cause adulteration.

		3. BMC to ensure that adulteration is avoided
4.	Post Harvest processing	1. Preliminary processing of the collected material, such as sorting and drying and stored in appropriate manner from pathogenic attack. 2. Basic post-harvest processes should be integrated at the local level.
	4.1 Traceability of raw materials	The collected materials are to be properly tagged with local and trade name, and source of collection.
	4.2 Minimizing post harvest wastage	Standards for this to be formulated.
5.	Trading society	1. Develop a MAP trading society with collection centres for the district. Society for medicinal plants collectors could enable better marketing and fair price and also address logistics problems. 2. No other way of resource mobilization permitted except for pharmacies cultivating medicinal plants. 3. Government should release specific directions that medicinal resources should be purchased by pharmaceuticals through LSG to prevent adulteration
	5.1 Information services	1. Rate to be fixed by BMC 2. Information services of prices of different species, demand & Market
6.	Laws and regulations	1. Mandatory registration of Collectors 2. All shops selling medicinal plants to be registered. This should be in line with the provisions of The Biodiversity Act (2002) that prior intimation to State Biodiversity Board for obtaining select biological resources for commercial utilization is necessary for corporate, associations or organizations. 3. Traditional rules and regulations for conservation in villages to be implemented through BMC together with other control measures, like provision of fine. 4. Industries to follow Good sourcing practices.
	6.1 GMP Certification and traceability of raw materials	1. Good Manufacturing Practices (GMP certification) should be linked with a provision that medicinal plant resources should be from sustainable and legal sources.
	6.2 Live Green label	1. The development of a Live Green Label for materials collected sustainably will provide an incentive for manufactures to collect raw materials sustainably. 2. This also ensures spurious drugs are avoided 3. Consumers should be made aware to choose products with the sustainable harvesting label
7.	Social responsibility and community development	1. Capacity building programs for collectors, and field staff. 2. Manufacturers and buyers should be encouraged to be a part of the Responsible Trade Network and exhibit their social responsibility
	7.1 Benefit sharing	3. Agreements with local communities to be based on knowledge of MAP resource value.
8.	Financial implications	1. Utilize Biodiversity fund prescribed for BMC 2. Revenue generated may be utilized for resource management

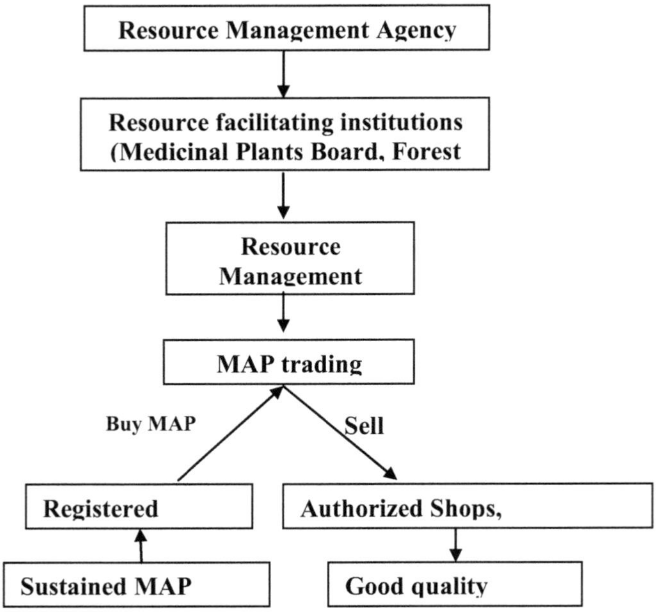

Figure 1: The schematic sketch of the structure of the action plan.

10.1. Anticipated Benefits of the Scheme

1. Conservation & sustained management of MAP.
2. Benefit sharing.
3. Local level management of MAP.
4. Unadulterated resources can be received from Trading society.
5. Collection according to availability.
6. Rate fixation by BMC.
7. Increased revenue for industries through systematic management of MAP resources in the wild.
8. Increased chances of livelihood and capacity development for local communities.
9. Existing traditional knowledge preserved.
10. Creation of awareness among the communities of joint local level management of natural resources.

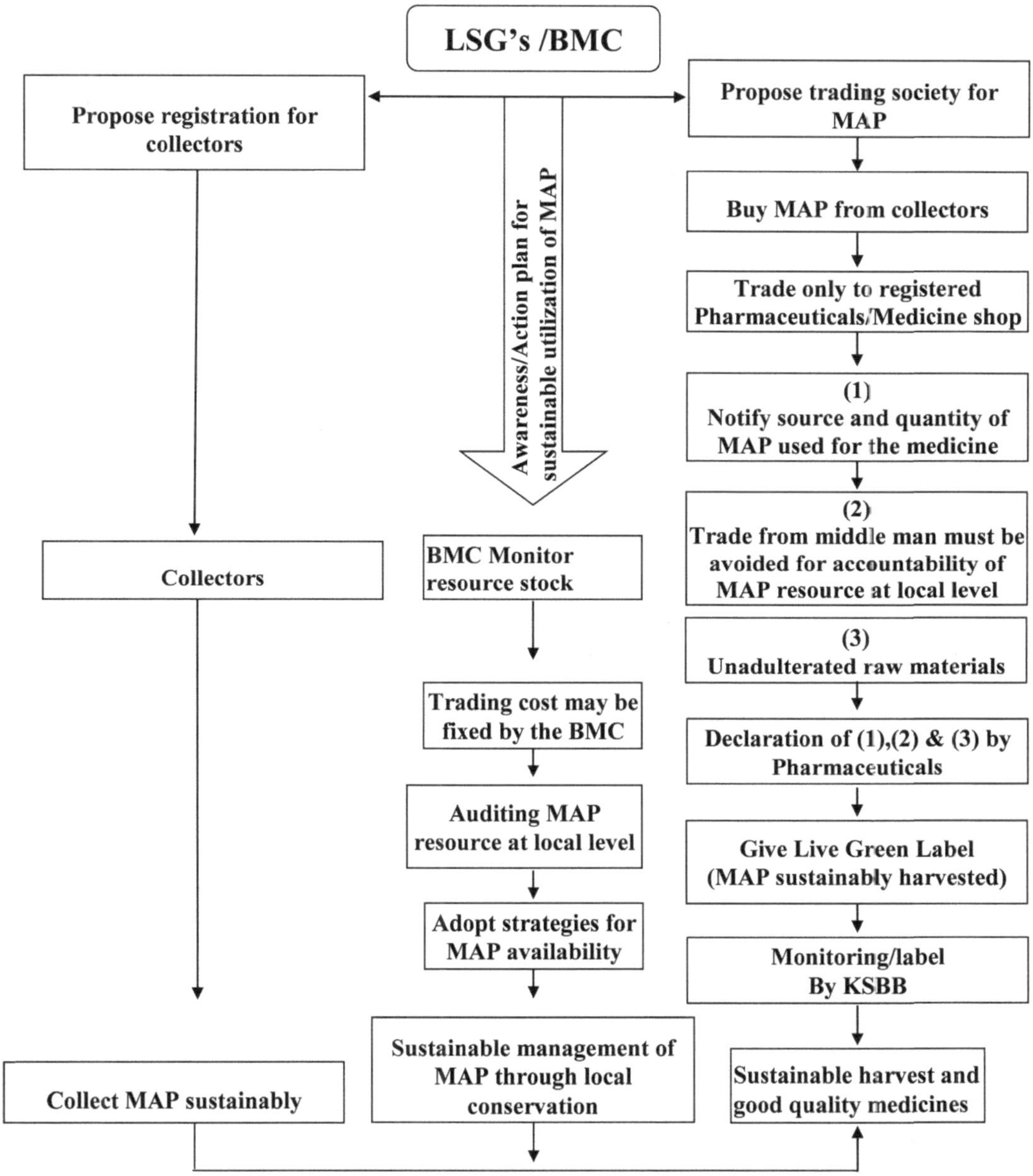

Figure 2: Detailed representation of proposed supply chain of MAP resource.

CONCLUSION

The purpose of the paper is to develop strategies for the conservation and sustained utilization of MAP species. A joint medicinal plan conservation and management approach is proposed involving all the stakeholders with good collection and sourcing practices combined with social responsibility and respect of traditional rights. Wild collection of MAP species shall be by proper inventory, and monitoring of harvesting impacts. Focus is on strengthening LSG & BMC for Resource Management and formation of medicinal plants trading society. Guidelines for conservation and capacity building for sustainable and non-destructive collection practices and controlling illegal harvesting of MAPs are detailed. Rare, endangered and threatened species and habitats likely to be affected by MAP collection are identified and protected. Local community's customary rights to sustainably collect MAP resources shall be respected.

Stricter enforcement of biodiversity legislations and mandatory registration for collectors from wild and herbal shops is recommended. The sustainable collection should be planned by BMC according to market requirements in order to minimize the wastage of products. The collected materials are to be properly tagged with local and trade name, and source of collection in order to support traceability to collection area. It is recommended that GMP certification should be linked with traceability of MAP and the development of Live Green label for materials harvested sustainably. Creation of a sense of ownership of natural resources among communities as a result of village based participatory management of resources will aid in conservation of biodiversity of MAP in the long term.

ACKNOWLEDGEMENTS

We are grateful to Kerala State Biodiversity Board and National Biodiversity Authority for providing necessary financial support.

CONFLICT OF INTEREST

The authors confirm that this chapter contents have no conflict of interest.

REFERENCES

[1] Bannerman RH, Burton J, Wen-chiieh C. Introduction in traditional medicine and health care coverage. WHO, Geneva 1983.

[2] Subrat N, Iyer M, Prasad R. The ayurvedic medicine industry: Current status and sustainability. Ecotech Services (India) Pvt. Ltd. and International Institute for Environment and Development 2002.

[3] Prasad R, Bhattacharya P. Sustainable harvesting of medicinal plant resources. In: Roy SB, Eds. Contemporary studies in natural resource management in India. New Delhi, India, Inter-India Publications 2003; pp. 168-198.

[4] FRLHT. Medicinal plants scenario in India. Report of the foundation for Revitalization of Local Health Traditions (FRLHT), Bangalore, India 2002.

[5] Tewari DN. Report of the Task Force on Conservation & Sustainable use of Medicinal Plants, Government of India, Planning Commission 2000.

[6] Subramanian MK, Nilakanta Pillai S, Thirumeni, Kannaiyan S. Biodiversity of Rice and its Conservation. National Biodiversity Authority, India 2007.

[7] www.keralaagriculture.gov.in: Conservation and development of medicinal plants. Accessed on 5/6/2012.

[8] FRLHT. Medicinal Plants of India: Guidelines for National Policy and Conservation Programmes. Foundation for Revitalization of Local Health Traditions (FRLHT), Bangalore, India 1997.

[9] Babu KL. Health and Livelihoods of Community and Traditional Medicinal Plants: SWOT of Two Agro climatic zones of India. Prepared by Center for Ecological Economics and Natural Resources. Bangalore Submitted to South Asia Network of Economic Research Institutes 2010.

[10] Rabinowitz D. Seven forms of rarity. In: Synge H, Eds. The biological aspects of rare plant conservation. John Wiley & Sons 1981; pp. 205-217.

[11] Durst PB, Mckenzie PJ, Brown CL, Appanah S. Challenges facing certification and eco-labeling of forest products in developing countries. International Forestry Review 2006; 8(2): 193-200.

[12] Lange D. Europe's medicinal and aromatic plants. Their use, trade and conservation. - Cambridge, UK, TRAFFIC International 1998.

[13] Robbins CS. American ginseng: the root of North America's medicinal herb trade. TRAFFIC, Washington 1998.

[14] Leaman DJ, Schippmann U, Glowka L. Environmental protection concerns of prospecting and producing plant-based drugs. In: Wozniak DA, Yuen S, Garrett M, Eds. International symposium on herbal medicine: a holistic approach: documents, proceedings and recommendations, 1-4 June 1997, Honolulu. International Institute Human Resources Development, San Diego 1997; pp. 352-378.

[15] Iqbal M. International trade in non-wood forest products: an overview. FAO, Rome 1993.

Chapter 3: The State of Frogs in India: A Case for Discontinuation of Use of Frogs in Dissections and Experiments

CHAPTER 3

The State of Frogs in India: A Case for Discontinuation of Use of Frogs in Dissections and Experiments

Mohammad A. Akbarsha*

Gandhi-Gruber-Doerenkamp Chair, Mahatma Gandhi-Doerenkamp Center (MGDC) for Use of Animals in Life Science Education, Bharathidasan University, Tiruchirappalli, Tamilnadu, India

Abstract: India has the pride of being one among the few mega-biodiversity countries of the world. Whereas its land area is limited to 2.4%, it has 7-8% of the total number of species so far recorded, of which about 45,500 species are plants and 91,000 species are animals. As reported in 2007, among the 5150 amphibian species recorded, 240 are present in India, which accounts for 4.66%. Unfortunately, the amphibians, which survived all the four mass extinctions since their origin, have been projected to be facing extinction at the global scale in the ongoing/envisaged sixth mass extinction. Many amphibians species have been declared in the Wildlife (Protection) Act 1972 as protected, and included in the different schedules reflecting different levels of threat. Frogs that are commonly used for dissection and experiments in India are included in Schedule IV of this Act. This means that holding of these frogs in captivity and/or use for any purpose including as delicacy without the permission from the local Chief Wild Life Warden is a punishable offence.

This happened in 1980, in an amendment to the Act of 1972. These frogs are caught for two major purposes, i) for human consumption as a delicacy; and ii) as a model organism in education and research. The Ministry of Environment and Forests, Government of India, and the University Grants Commission have banned use of these frogs in any aspects of education. But frogs continue to be used in Zoology and Pharmacology laboratory exercises. This needs to be stopped so as to conserve frog biodiversity in India.

Keywords: Amphibians, biodiversity, curriculum, dissections, extinction, frogs, habitat degradation, infections, overharvesting, penalty, pharmacology, physiology, pollution.

***Corresponding author Mohammad A. Akbarsha:** Gandhi-Gruber-Doerenkamp Chair, Mahatma Gandhi-Doerenkamp Center (MGDC) for Use of Animals in Life Science Education, Bharathidasan University, Tiruchirappalli 620024, Tamilnadu, India; Tel: +914312420557; Mob: +919790995854;
E-mails: mgdcaua@yahoo.in; director@mgdccloud.org

Laladhas K.P., Oommen V. Oommen and Sudhakaran P.R. (Eds.)

1. INTRODUCTION

It is argued that we are going through the sixth great mass extinction [1, 2]. The first five mass extinctions are the ones during the transition from Ordovician to Silurian (˜439 Mya), the Late Devonian (˜364 Mya), the Permian-Triassic (˜251 Mya), the End Triassic (˜199-214 Mya), and the Cretaceous-Tertiary boundary (˜65 Mya) [3]. The amphibians are the most affected animals during the ongoing mass extinction.

Though many of the amphibian species perished during the four mass extinctions since their first evolution, representatives of all three orders survived to date [2]. There is a "general understanding that a larger proportion of amphibian species are at risk of extinction" now [4]. In the recent past, they have been disappearing at a faster rate than ever [2]. The International Union for Conservation of Nature (IUCN) has observed that "as many as one third of the nearly 6,000 known amphibian species— frogs, toads, salamanders, worm-like caecilians— are threatened with extinction" [4]. The "leading causes of biodiversity loss in general, and all of these acting alone or together are responsible for modern amphibian declines: commercial use; introduced/exotic species that compete with, prey on, and parasitize native frogs and salamanders; land use change" (habitat modification and destruction); "contaminants" (pesticides, fertilizers and byproducts of human activities); "climate change" visa vi global warming; and "infectious disease" [2, 4, 5]. Thus, "multiple factors acting synergistically are contributing to the loss of amphibians. Behind all of these activities is one weedy species, *Homo sapiens*, which has unwittingly achieved the ability to directly affect its own fate and that of most of the other species on this planet. "It is an intelligent species that potentially has the capability of exercising necessary controls on the direction, speed, and intensity of factors related to the extinction crisis" [1]. Another way this weedy species is threatening the existence of frogs and toads is over-harvesting, which is generally "a less pervasive threat than habitat loss, but it is more likely to drive a species into rapid decline" [6]. Consumption as food is one aspect of overharvesting which is biological, an aspect of the food web, but use in experiments is now added to the list of threats to amphibians, particularly as millions of frogs are used in scientific research or classroom experiments which are non-biological. In India, frogs are killed in such large numbers for the purpose of education and experiments, that populations of these animals have been driven to the edge, and they are no more available in such a large number as they used to be. The chapter presents a review of the status of these animals based on the author's personal observation and experience.

2. THE STATE OF AMPHIBIANS IN INDIA

India is proud as a mega-biodiversity country. Whereas its land area is limited to 2.4%, it has 7-8% of the total number of species so far recorded of which about 45,500 species are plants and about 91,000 species are animals. As reported in 2007, among the 5150 amphibian species recorded, 240 are present in India, which accounts for 4.66% [7]. Also, new species are being reported, which are heartening. Unfortunately, the amphibians which survived all the four mass extinctions since their origin have been projected to be facing extinction at the global scale, as an aspect of the ongoing/envisaged sixth mass extinction [1], and India is no exception. "India has the third largest amphibian population in Asia. The amphibian fauna of India comprises of 272 species of which 167 (66.3%) are endemic to the country. In spite of its broad variety of species, India holds second place on the list of countries having the most number of threatened amphibian species in Asia, with 67 (25%) of its species facing possible extinction. Out of the 38 species of amphibians in Asia that are confirmed to be extinct, by the Global Amphibian Assessment (GAA), 1 is from India. In addition, 13 species are listed as critically endangered, 31 as endangered and 23 as vulnerable. Further, 95 species are listed under the data deficient category indicating that the number of threatened species may be much higher once information becomes available" (http://www.iucn.org). The causes of the extinction in the anvil are many but all are man-made and most are environmental, essentially climate change, global warming, habitat degradation, pesticide pollution, and infections [2, 8]. In this backdrop, it is highly relevant that the amphibians of our country, the rich heritage for our future generations, are conserved.

3. FROGS BELONGING TO THE GENUS RANA ARE COVERED BY THE WILDLIFE (PROTECTION) ACT OF INDIA

Many species of amphibians have been declared in the Wildlife (Protection) Act 1972 as protected, and included in the different Schedules reflecting different levels of threat. Frogs that are commonly used for dissection and experiments, *Rana spp*, in India are included in Schedule IV of this Act. This means that holding of these frogs in captivity and/or use for any purpose without permission from the local Chief Wild Life Warden is a punishable offence (www.moef.nic.in/legis/wildlife/wildlife2s4.html;www.bankuraforest.in/). This happened in 1980, in an amendment to the Act of 1972.

4. OVERHARVESTING OF FROGS IN INDIA

These frogs are caught for two major purposes, i) for human consumption as a delicacy [9]; and ii) as a model organism in education and research [10, 11]. Millions of tones of hind legs harvested from a few selected species have locally decimated frog populations [10]. In India there was an immense pressure on handful of species including *Hoplobatrachus tigerinus* (Tiger frog), *Hoplobatrachus crassus* (Jerdon's bullfrog), *Euphlyctis hexadactylus* (Indian green frog or Indian pond frog) and *Euphlyctis cyanophlyctis* (Indian skipper frog). As much as 10,000 tons of frogs were harvested from India in 1983 to meet the international culinary demands, which accounts for sacrifice of about 40 to 100 million frogs in 1983. Such unrestrained harvest grossly reduced the number of adults belonging to the four large edible frog species. Export of frog legs has been banned in 1987 and capture from the wild for the purpose of consumption has greatly reduced. The four species of frogs mentioned above and a few others, including *Bufo melanostictus* and *Limnonectes limnocharis*, are being collected in a large number to meet the demands of Indian students of Biology. Starting at high school and through undergraduate Zoology education an average Indian student of Biology is required to sacrifice at least 30 frogs as part of his/her practical lesson. Frogs are used by medical and paramedical students for subjects such as pharmacology and physiology. Furthermore, the present author would put the number of frogs sacrificed by each student at a modest average of 5. India has at any one time, over 2 million undergraduate and graduate students who take courses in science and require animal dissection/experiments. At the rate of 5 frogs per head, the number would be 10 million per annum. The Central Board of Secondary Education (CBSE) Senior Secondary had about 2 million students dissecting/experimenting with frogs and thus used another 10 million frogs per annum. With the launching of 10 + 2 + 3 pattern of education in 1979, the +2 came to be a stand-alone stream and each student taking to biology/zoology subject had to dissect not less than 5 frogs. These students, in turn, used about 80 million frogs per annum. Thus, approximately 100 million frogs were caught from the wild and sacrificed each year. This overharvesting of frogs for human consumption and education became a compounding factor to the other causes, rendering depletion of frogs in the wild.

5. DISSECTION OF/EXPERIMENTS USING FROGS GETTING DELETED FROM CURRICULUM

Due to this depletion, frogs were not available for supply, which forced the regulatory authorities to remove at large dissection of and experiments using frogs at the CBSE Senior Secondary and State Boards +2 level. One of the reasons for the decision of the regulatory authorities has been the pressure mounted by the animal welfare organizations, and there were court cases too. Thus, the CBSE deleted all exercises relating to animals in biology practical in the senior school curriculum in 2001. The Higher Secondary/Intermediate Education Boards of most of the States in India also dropped exercises involving animals from the curriculum. The latest to join this trend is Andhra Pradesh, and the decision is revealed in the curriculum revised during 2012-13. But use of frogs in laboratory exercises in zoology and pharmacology is continuing, in spite of the notices from the Union Government Ministry of Environment and Forests (MoEF) and the University Grants Commission (UGC), New Delhi, the Regulatory Authority of higher education in India. The number of frogs thus caught from the wild and sacrificed cannot be underestimated [12]. These tentative estimates apart, it is necessary that scientific investigations on the population dynamics and decline are addressed, but inclusion of *Rana* spp of frogs by IUCN in the Red List is authoritative enough as substantiation of decline of these frogs.

6. TEACHERS CONTINUE TO INSIST ON USE OF FROGS IN DISSECTIONS AND EXPERIMENTS

Many teachers, belonging to these domains, who have inherited the legacy from their own teachers, under conditions very different then, have been very sensitive and emotional that use of frogs in dissection and experiments should continue [13] and to them any attempt to stop this practice will mean killing zoology/pharmacology [14]. All these are in the backdrop of provisions in the Constitution of India that safeguard the interest of living creatures. Section 51A (G) of the Constitution states: "it shall be the duty of every citizen of India to protect and improve natural environment including forests, lakes, rivers and wildlife, and to have compassion for all living creatures". As provided for in the Prevention of Cruelty to Animals Act, 1960, Sec 17.2 (d): "Experiments on animals are avoided wherever it is possible to do so; as for example: in medical schools, hospitals, colleges and the like, if other teaching devices such as books, models, films and the like, may equally suffice", and in Sec 17.2 (f): "As far as possible, experiments on animals are not performed merely for the purpose of acquiring manual skill".

7. OVERHARVESTING OF FROGS AS CAUSE OF HABITAT DISPLACEMENT

In the experience of this author, during 1970's the Indian green frog, then *Rana hexadactyla*, and now *Euphlyctis hexadactylus*, was available in plenty. For teacher training undergraduate students in dissections, about 150 frogs were used each day. As years passed by, the supply was in problem. The specimen suppliers supplied a smaller version of frog which by colour and size was not *E. hexadactylus*. Today *E. hexadactylus* is almost absent in many places in Southern India, and the most common frog available is *Euphlyctis cyanophlyctis*, which was rare in those days in Tamil Nadu but abundant in the northern parts of peninsular India. This, in core ecological science, is known as habitat displacement, which impacts biodiversity and ecological niche.

8. THE UGC GUIDELINES FORBID USE OF FROGS FOR CLASSROOM PURPOSES

Therefore, the University Grants Commission, New Delhi, the apex regulatory body of higher education in India (UGC), in the Guidelines to phase out animal dissections from Zoology/Life Science curriculum November 2011, has totally banned use of the protected animals for dissections and experiments. Thus, frogs under the genus *Rana*, already threatened and included in the IUCN Red List, though not under the category "critically endangered" [15], can not at all be used in dissections and experiments unless with permission from Competent Authorities (invariably, the Authorities deny permission, given the alarming threat to these species in India). Yet, there is still a section of teachers who engage the students in dissection of frogs in undergraduate basic sciences, and experiments using frogs in Physiology and Pharmacology.

9. A ROLE FOR MGDC IN CURTAILING USE OF FROGS IN DISSECTIONS AND EXPERIMENTS

It was suggested that frogs must be bred in captivity, and the list of such breeders be made available for teachers to procure the frogs thus bred for use in the laboratory [8], but it is sad that all attempts to breed Indian frog species in captivity have not yielded good results, and if there is claim that frogs bred in captivity is available for sale it must be a dubious claim and the animal should have been captured from the wild [13]. The Mahatma Gandhi-Doerenkamp Centre (MGDC), in collaboration with some enlightened teachers and a few voluntary organizations (PETA-India, People for Animals, International Centre for Alternatives in Research and Experiments), are doing their best to sensitize these teachers not to continue this practice. This is a service to help conservation of the frog species. In doing this sensitization this issue is viewed from the perspectives of course content, teaching methods, ecological imbalance that will be caused, legal provisions and animal ethics. The teachers are encouraged to use computer-assisted learning (CAL) in the place of dissections and animal experiments [16-20].

10. NAME OF FROGS IN THE WILDLIFE (PROTECTION) ACT AND THE NAMES CHANGED THEREAFTER

A note of importance here is that since 1980, when freshwater frogs were brought under the purview of Wildlife (Protection) Act, many species have been removed from Fam: Ranidae and so their names are different now- for example *Rana hexadactyla*, the green frog- the most preferred model animal for dissections and experiments, is now *Euphlyctis hexadactylus*; *Rana tigrina* is now *Hoplobatrachus tigerinus*; *Rana breviceps* is now *Sphaerotheca breviceps*; *Rana crassa* is now *Hoplobatrachus crassus*; and *Rana cyanophlyctis* is now *Euphlyctis cyanophlyctis*- and all have been moved to Fam: Dicroglossidae. This might provide leverage for those who still harvest these frogs for dissections and/or experiments. It is important that these changes in the nomenclature and taxonomic status of these important frog species are adopted in the Wildlife (Protection) Act (1972) and in the Guidelines/circulars that are brought up from time to time.

11. THE STATUS OF CULTURE OF FROGS AT COMMERCIAL SCALE

It is generally argued that culture of frogs at a commercial scale will be a solution to this problem. There have been quite a few studies aiming at the general biology of the common species of frogs, with special reference to ecology, food and feeding, reproductive biology and life cycle. Here, the emphasis has been to overcome the ban on capture of frogs from the wild for purpose of consumption as a delicacy. Considered lucrative, there have been attempts in this direction but the venture has not been a success. *Rana catesbeiana*, the bull frog which weighs 450-500 g has been claimed as successfully cultivated to commercial scale, not in India but elsewhere but not yet as an agro-industry [www.thaiscience.info/Article for Thai Science/Article/2/Ts-3 frog farming in Thailand.pdf]. Even here the claim is disputed [http://mdc.mo.gov/sites/default/files/resources/2010/05/4899_2850.pdf]. "Those few individuals who claim to be successful frog farmers generally are distributors engaged in the selling of adult frogs, tadpoles, or frog eggs, frequently harvested from the wild [http://pubs.ext.vt.edu/420/420-255/420-255_pdf.pdf]. The causes underlying this failure have been many, such as long time two years for full-sized frog to harvest, and the consequent economic unfeasibility, dependence of breeding on environmental factors, *etc*. Scientific reports on the possibility of frog culture in India are few, and these studies attribute difficulties especially the pesticide contamination of the pond water [21,

22]. As things stand now, farmed frogs are not going to be a reality in the light of several complexities.

CONCLUSION

The abundance of the various species of frogs is not documented appropriately since in the opinion of this author the picture concerning *E. hexadactylus* is not that bright as projected in a few recent publications. This is highly pertinent since the NBAP has envisaged as one of the objectives "Development and integration of biodiversity Databases- to collect, collate and integrate biodiversity information from diverse sources into a national database on different components of biodiversity with distributed networking systems and linkages, and to intensify the survey, identification and inventorization of country's floristic, faunal and microbial resources with special attention to hitherto unexplored areas, and keystone, umbrella, endangered and endemic species which need to be conserved on priority basis" [nbaindia.org/uploaded/Biodiversityindia/NBAP.pdf]. The loss of biodiversity and the consequent ecological imbalance is more important.

ACKNOWLEDGEMENTS

Declared None.

CONFLICT OF INTEREST

The author confirms that this chapter contents have no conflict of interest.

REFERENCES

[1] Wake DB, Vredenburg VT. Are we in the midst of the sixth mass extinction? A view from the world of amphibians. Proc Natl Acad Sci 2008; 105(1): 11466-73.

[2] Collins JP. Amphibian decline and extinction: what we know and what we need to learn. Dis Aquat Organ 2010; 92: 93-9.

[3] Erwin DH. Lessons from the past: Biotic recoveries from mass extinctions. Proc Natl Acad Sci USA 2001; 98: 1399-403.

[4] Stuart S, Chanson JS, Cox NA, Young BE, Rodrigues ASL, Fischman DL, Waller RW. Status and trends of amphibian declines and extinction worldwide. Science 2004; 306: 1783-6.

[5] Daszak P, Cunningham AA, Hyatt AD. Infectious disease and amphibian population decline. Diversity Distributions 2003; 9: 141-50.

[6] Xie F, Lau MW, Stuart SN, Chanson JS, Cox NA, Fischman DL. Conservation needs of amphibians in China: A review. Sci China C Life Sci 2007; 50: 265-76.

[7] India's Fourth National Report to the Convention on Biological Diversity. Ministry of Environment and Forests, Govt of India, New Delhi 2009.

[8] NBAP. National Biodiversity Action Plan. Ministry of Environment and Forests, Government of India 2008.

[9] Nayak Y. Why ban use of frog in laboratory? Curr Sci 2010; 98: 1162.

[10] Seshadri KS, Vivek Chandran A, Gururaja KV. Anurans from wetlands of Puducherry, along the East Coast of India. Check List 2012; 8: 23-6.
[11] Daniels RJR. Amphibians of Peinnsular India. Universities Press (India) Private Limited. Hyderabad, India 2005; pp.35-45.
[12] Akbarsha MA. Movement to curtail animal dissections in Zoology curriculum: Review of the Indian experience. ALTEX 2007; 24: 163-6.
[13] Subramoniam A, Gitanjali B. CPCSEA- A double-edged sword. Indian J Pharmacol 2003; 35: 71-2.
[14] Shehnaz SI, Sreedharan J, Mathew E, Gomathi KG, Sami Khan N. Willingness to spare animals in undergraduate medical education in Southern India: a preliminary questionnaire-based investigation. Altern Lab Anim 2011; 39: 557-66.
[15] Venkatesha MG. Is dissection of animals in Zoology redundant?.Curr Sci 2007; 92: 568.
[16] Sarkar J. Critically endangered Indian animals. Curr Sci 2011; 100: 1608-9.
[17] Akbarsha MA, Zeeshan Md, Pereira S. CPCSEA National Conference, India- Report. ALTEX 2011; 28: 153-5.
[18] Akbarsha MA, Zeeshan Md and Meenakumari KJ. Alternatives to animals in education, research and risk assessment: An overview with special reference to Indian context. ALTEX Proceedings 2013; 2/13: 5-19.
[19] Raveendran R, Gitanjali B. Alternatives to Animals in Teaching: Experience in an Indian Medical School. ALTEX Proceedings 2012; 1/12: 355-60.
[20] Sathyanarayana MC. Need for alternatives for animals in education and the alternative resources. ALTEX Proceedings 2013; 2/13: 77-81.
[21] Sharma KK. Modern tools of species identification can save millions of animals killed in identification and animal taxonomy. ALTEX Proceedings 2013; 2/13: 65-8.
[21] Marian MP, Arul V, Pandian TJ. Acute and chronic effects of carbaryl on survival, growth and metamorphosis in the bull frog (*Rana tigrina*). Arch Environ Contam Toxicol 1983; 12:271-275.
[22] Pandian TJ, Marian MP. Production and utilization of frogs: an ecological view. Proc Indian Acad Sci (Animal Sci)1986; 95:289-301.

APPENDIX

Elusive and "Lost" Caecilian Species of Western Ghats, India - A case study

Ramachandran Kotharambath[1,*], Mark Wilkinson[2], Lekha Divya[3], Oommen V. Oommen[4], Sanil George[5], Ronald A. Nussbaum[6] and David J. Gower[2]

[1]Department of Zoology, Government College Chittur, Palakkad, Kerala, India.678104; [2]Department of Zoology, The Natural History Museum, London, SW7 5BD, UK; [3]Department of Animal Sciences, School of Biological Science, Central University of Kerala, Riverside campus, Padanakadu, Kasaragod, India; [4]Kerala State Biodiversity Board, L-14 Jainagar, Thiruvananthapuram 695011, Kerala, India; [5]Molecular Medicine Laboratory, Rajiv Gandhi Centre for Biotechnology, Thiruvananthapuram, Kerala, India and [6]Museum of Zoology and Department of Ecology and Evolutionary Biology, University of Michigan, Ann Arbor, MI 48109-1079, USA

Abstract: The taxonomy of most of the Indian caecilian species is unstable and available knowledge is based on a few specimens collected mostly from disturbed or anthropogenic habitats. Some species have not been recorded for decades since their description, resulting in a poor understanding on this group of carnivorous limbless subterranean amphibians. Among the Western Ghats striped Ichthyophiids, which are among the most conspicuous and colourful caecilian species, *Ichthyophis longicephalus* Pillai (1986) had been the least known for almost two and half decades after its discovery from the Silent Valley forests. The taxonomic, systematic and conservation status of *I. longicephalus* has been reassessed [1]. Here we briefly describe the background efforts that went in to the rediscovery of this species, mention the implications of poor specimen preservation in caecilian taxonomy and present some supplementary data.

Keywords: Amphibians, Caecilian, diversity, *Ichthyophis longicephalus,* systematics, taxonomy, Western Ghats.

1. SEARCH FOR THE "LOST AMPHIBIANS"

During the monsoon months of 2010, Conservation International (CI) and the IUCN SSC Amphibian Specialist Group (ASG), with support from Global Wildlife Conservation (GWC), launched "Search for Lost Amphibians" to document the survival status of those amphibian species which had not been recorded for more than a decade. This five-month search involved 126

***Corresponding author Ramachandran Kotharambath:** Department of Zoology, Government College Chittur, Palakkad, Kerala, India; E-mail: ramachandrank6@gmail.com researchers in 21 countries looking for mainly "lost" frog species. Of the two caecilian teams in this programme one (and the only one from India) was for *Ichthyophis longicephalus* Pillai. More than 255 hours of search at 11 localities resulted in seven new specimens of *I. longicephalus* but the 47 hours of search in the Silent Valley forests, the holotype locality, did not yield any specimens.

2. TAXONOMY AND SYSTEMATICS

We could refer the new specimens to *Ichthyophis longicephalus* based on careful and patient morphological and molecular studies, carried out over a period of almost three years, of the new material and also extracting every bit of usable data from the heavily damaged holotype specimen [1]. The study would have taken less time had the holotype been in good shape. For a species with subterranean lifestyle and very few people working on them the damage to the type specimens can have serious implications for the taxonomy of the species (and the larger group it belongs) in the long run. Researchers could well be forced to keep aside projects on species with poorly maintained types and unreliable and unspecific locality data. Here in the case of *I. longicephalus* the extensive and focused field and lab works helped document the current survival status and contributed substantially towards the taxonomic understanding of the species.

Figure 1: A-B. Photographs of the new (2009-2010) Ichthyophis longicephalus specimens in life from Kalarikkal.

Figure 2: A-B. Photographs of the new (2009-2010) Ichthyophis longicephalus specimens from Aralam Wildlife Sanctuary.

3. ISSUES IN CONSERVATION OF DIVERSITY OF *CAECILIAN* SPECIES

The taxonomy of most of the Indian caecilian species needs to be stabilized and/or revised if we have to have a solid understanding about their diversity, population and conservation status. The last decade has seen many new caecilian species descriptions but not many works on cementing the basic taxonomy of the species which have not been recorded reliably for a long period. The case of *I. longicephalus* throws light on the damages improper specimen maintenance can have on taxonomy of an inconspicuous species.

ACKNOWLEDGEMENTS

Declared none.

REFERENCE

[1] Kotharambath R, Wilkinson M, Oommen OV, George S, Nussbaum RA, Gower DJ. Onthe systematics, distribution and conservation status of Ichthyophis longicephalus Pillai, (Amphibia: Gymnophiona: Ichthyophiidae). Journal of Natural History 1986; 46:47-48, 2935-2959.

CHAPTER 4

The Role of People's Participation in the Conservation and Management of Biodiversity and Associated Traditional Knowledge

Rajasekharan S.[*]

Division of Ethnomedicine and Ethnopharmacology, Jawaharlal Nehru Tropical Botanic Garden and Research Institute (JNTBGRI), Palode, Thiruvananthapuram, Kerala, India

Abstract: The chapter highlights the ancient and modern concepts/strategies of biodiversity conservation through people's participation both at *in-situ* and *ex-situ* level and also focuses on the significance and sector wise thrust areas of Traditional Knowledge (TK) including their protection. A development paradigm, which is not exclusively in the hands of the professionals and technologists, but with the local people should be developed. The need for systematic documentation of TK using prior informed consent, legal protection of TK, database inventory (defensive protection), exclusive rights of the beneficiaries, complementary protective measures and protection by signing contractual agreement with knowledge holders/providers and future strategies and action plan to be implemented through the people's participation in management of biodiversity are discussed in detail.

Keywords: Ayurveda, biodiversity conservation, botanic garden, DNA library, gene bank, pollen bank, sacred groves, traditional knowledge.

1. INTRODUCTION

Biodiversity is the life line of planet Earth and provides a wide range of ecosystem services necessary for the livelihood and well-being of people. According to the Millennium Ecosystem Assessment, the number of species on earth ranges from 5 to 30 million and out of this only 1.7 - 2 million species has been formally identified. 'India is one among the 17 mega diverse countries [1]". "It covers only 2.4 % of the land area, but accounts for 7-8% of the recorded species of the world, spread over 45,500 species of plants and 91,000 species of animals" [1].

For implementing conservation programmes, it has to be ensured that the end result should be a development paradigm, which is not exclusively in the hands of the professionals and technologists, but with the local people.

***Corresponding author Rajasekharan S.:** Division of Ethnomedicine and Ethnopharmacology, Jawaharlal Nehru Tropical Botanic Garden and Research Institute (JNTBGRI), Palode, Thiruvananthapuram, Kerala, India; Tel: 9446557914; E-mail: drrajsek@yahoo.com

Laladhas K.P., Oommen V. Oommen and Sudhakaran P.R. (Eds.)

This paper, highlights the ancient and modern concepts/strategies of biodiversity conservation through people's participation both at *in-situ* and *ex-situ* level and also focuses on the significance, sector wise thrust areas of Traditional knowledge including the necessity of protecting such knowledge for the overall welfare of human beings and other species on earth.

2. ANCIENT INDIAN CONCEPTS/STRATEGIES OF BIODIVERSITY CONSERVATION

Ayurveda is a *'Sastra'* different from Modern medicine and denotes the 'Science of life', the word 'Ayu 'means' Life' and 'Veda' means 'Knowledge'. For information relating to plants or animals, the terms 'Vrikshayurveda'(Ayurvedic arboreal medicine) and 'Mrigayurveda' (Ayurvedic veterinary medicine) are used respectively [2]. Ayurveda identifies man as an integral part of nature and stresses the necessity of maintaining complete harmony with all living and non-living components of the surroundings (Environment).

Major subjects brought under Vrikshayurveda include *Bhoominiroopanam* (identification and classification of soil conservation technique), *Beejoptividhi* (sowing techniques), *Padapavipaksha* (method of propagation), *Ropanavidhanam* (plantation technique), *Nishechanavidhi* (irrigation technique), *Poshanavidhi* (special nutritional care and management of plants with various biotechnological means), *Drumareksha* (conservation practices), *Taruchikitsa* (treatment with biomedicine), *Nivasasannatharusubha- asubhalakshanam* (selection of plants for planting near the residence based on their good/bad qualities.), *Tarumahima* (glory of trees – awareness programme) *Upavanakriya* (landscaping and gardening) and *Chithreekaranam* (research on plant biotechnology).

While discussing the conservation strategies of the aquatic and terrestrial biodiversity, Vrikshayurvedic scholars have highlighted a number of *in-situ* and *ex-situ* conservation programme with people's participation. Importance of *ex-situ* conservation of tree species with high carbon intake properties is specifically recommended for planting to avoid the negative impact of climatic change. Therefore it is suggested that, one individual has to plant following plant species and also construct a pond nearby during his/her lifespan.

"Aswathamekhampichumandamekham, Nygrodhamekhamdesathintrinika,

Kapithavilwaamalakatrayamcha, Panchambravapinarakamnapasyet" [3]

Table 1: Name of the Plant species recommended

Botanical Name	Local Name	Number of the Plant species to be planted
Ficus religiosa L.	Arayal	One
Azadirachta indica A. Juss.	Aryaveppu	One
Ficus benghalensis L.	Peral	One
Flacourtia Montana Graham	Charalmaram	Three
Aegle marmelos Corr.	Koovalam	Three
Phyllanthus emblica L.	Nelli	Three
Mangifera indica L.	Maavu	Five
Tamarindus indica L.	Vaalanpuli	Ten
Total Number		27

2.1. BIODIVERSITY CONSERVATION PROGRAMME WITH COMMUNITY PARTICIPATION

2.1.1. Nakshathra Vrikshas (Star Trees)

This concept was developed by the Vrikshayurvedic scholars for conservation of biodiversity. To implement this concept, they developed a holistic strategy for people's participation by establishing connectivity between stars, plants, human beings, animals and birds. Their vision was to establish an *ex-situ* conservation site with 27 trees which possess high therapeutic and aesthetic value. Every individual is advised to plant the tree governing his/her star. This information is based on Vrikshayurveda. The tree mentioned against each star is; Aswathi (*Strychosnux-vomica* –Kanjiram), Bharani (*Phyllanthus emblica* – Nelli), Karthika (*Ficus racemosa* - Atthi), Rohini (*Syzigium cumini* - Njaval), Makayirom (*Acacia chundra* - Karingali) Thiruvathira (*Gmelina arborea* - Kumbil), Punartham (*Bambusa bambos* - mula), Pooyam (*Ficusre ligiosa* - Arayal), Aayilyam (*Mesuaferrea* - Nagapoo), Makom (*Ficus benghalensis* - Peral), Pooram (*Butea monosperma* - Chamata), Uthram (*Ficus microcarpa* - Ithi), Attham (*Spondias pinnata* - Ambazham), Chittira (*Aegle marmelos* - Koovalam), Chothi (*Terminalia arjuna* – Neermaruthu), Visakom (*Flacourtia montana* - Vayamkatha), Anizham (*Mimusops elengi* - Elanji), Ketta (*Symplocos cochinchinensis* - Pachotti), Moolam (*Vateria indica* – Veluthakunthirikom), Pooradam (*Barringtonia racemosa* – Samudrakai), Uthradam (*Artocarpus heterophyllus* - Pilavu), Thiruvonam (*Calotropis gigantea* - Erukku), Avittom (*Prosopis cineraria* - Paarampu), Chatayam (*Neolamarckiacadamba* -

Kadambam), Pooruruttati (*Mangifera indica* - Maavu), Uttrattathi (*Azardirachta indica* - Veppu), Revathi (*Madhuca neriifolia* - Iluppa) [4].

2.1.2. Sacred Groves

At present there are 2,000 well preserved scared groves in Kerala. This includes Sarppakavu, Ayyappankavu, Devakavu, Pavithravana, Yekshikavu, sacred groves associated with temples and Tharavadu. The presence of 15,000 sacred groves in Travancore and Cochin States alone has been recorded.

It is also noted that conservation of different varieties of snakes in Kerala was a practice closely linked with cultural heritage. Therefore, majority of 'Kavus' (Sacred groves) were Sarpakavu (Serpent groves). Some of the families including Brahmins are still preserving the Sarpakavu in their vicinity. Sacred groves maintained by the Mannarasala temple near Harippad in Alappuzha district, Kerala is a shrine were snake worship has been sanctified for centuries.

2.1.3. Tribal Concept of Biodiversity Conservation

As per the historical records, Malapandaram tribes settled in Achankovil area of Kollam district of Kerala are the descendents of nomadic primitive hunter gatherers who once roamed the hill tops and caves. According to the tribes, they are the 'Autochthons' of the Kerala forest. They preserved forest like their own property, and maintained the tradition of fully protecting certain pockets of dense evergreen forest located along the perennial streams popularly known as '*Kanakom*'. '*Kanakom*' was an area exclusively preserved by the tribes as a habitat for the animals, plants and other living beings without any disturbances. The tribes never entered this area and fully protected this part as location specific 'Hot spots'. The tribes also established a symbiotic relationship with the forest and used the forest wealth in a sustainable manner for their healthy survival [5].

3. MODERN CONCEPT OF BIODIVERSITY CONSERVATION

3.1. *In situ* Conservation

Establishment of Biosphere Reserve, National Park, Wildlife Sanctuaries, Sacred groves and other protected areas are well-established methods of conservation.

Pilot participatory programme on conservation and sustainable utilization of Medicinal and aromatic plants under the Kerala forestry programme aided by World Bank has been implemented by Jawaharlal Nehru Tropical Botanic Garden and Research Institute (JNTBGRI). This participatory programme consists of 9 major activities for empowering the tribal people with a view to improve the quality of their lives linking local biodiversity and indigenous knowledge.. The nine activities include, contact and general awareness programme, mapping the general profile of the tribal settlement and adjoining forest area, conservation and cultivation of medicinal and aromatic plants, value addition and chemical prospecting, production of value added products and its utilization, training programme on post harvesting technology, entrepreneurship development and establishment of Co-operative societies/self-help groups and marketing strategies. In view of the merit of the programme, as per the recommendation of IUCN, Srilanka, 3 groups of community leaders consisting of 15-18 members visited the implementation site (Kani Tribal settlements, Kottor, Thiruvananthapuram) as part of their familiarization programme.

Another novel attempt to conserve biodiversity - rich unique forest ecosystem with the participation of tribal people was the 'Development of Gene pool Conservation Areas' under the Western Ghats development programme of Kerala Forest and Wild life Department with the support of Vanasamrakshnasamithy.

3.2. *Ex situ* Conservation

Conserving the plants outside the natural habitat would lead to long term conservation. India has around 140 botanical gardens, including those affiliated to Universities. The field gene bank programme launched by TBGRI from 1992 to 1999 is now acclaimed as a very effective method of conservation of medicinal and aromatic plant genetic resources. 'Herbs for all and Health for all' is one of the people's participation programmes on conservation, cultivation and sustainable utilization of medicinal and food plants, successfully implemented at Vithura Gramapanchayath of Thiruvananthapuram district under the *ex-situ* conservation activities of JNTBGRI.

4. FUTURE STRATEGIES AND ACTION PLAN FOR BIODIVERSITY CONSERVATION

- Identification of Gene pool conservation area in the forest ecosystem with a view to protect medicinal and food plants including rare, endangered and endemic medicinal plants which are likely to face the impact of climate change.

- Establishment of Gene Pool Conservation Area (GPCA) of medicinal and food plants in the forest eco-system.

- Introduce various measures to reduce the forest dependency of ecosystem people.
- Documentation of traditional knowledge related to plants used for food and medicinal purpose and its IPR protection and benefit sharing.
- Establishment of nurseries for selected medicinal plants including rare, endangered and endemic.
- Establishment of new network for the collection of NWFP. It includes introduction of good collection, processing, storage and selling practice including value addition.

5. TRADITIONAL KNOWLEDGE

TK is restricted to location specific knowledge of common people including ethnic communities residing in a particular region/country, the knowledge is confined to the genetic and non-genetic resources available within their surroundings.

5.1. Significance of TK

- TK is directly linked with the cultural heritage of the people and has been handed over from generation to generation.
- TK is getting eroded rapidly because of the changing life styles of the people; hence there is an urgent need to document such valuable information.
- To evaluate medicinal and food value of plants, the role of TK is important for developing diverse novel products.
- Technology transfer of these products will further ensure their commercialization and benefit sharing with the knowledge providers [6].

5.2. Necessity for Systematic Documentation

Systematic documentation is essentially required for;

- Establishment of a network to access the data by the end users.

- Research and Development activities, Product development, Patenting, Technology transfer, Commercialization and Benefit Sharing.
- Integration of TK in to planning process for the various developmental activities at grass root level.
- Formulation of new protective measures of TK and implementation of effective legal frame work to check bio-piracy.

5.3. Protection of Traditional Knowledge

5.3.1. International Initiatives

CBD, signed at the United Nations Conference on Environment and Development (UNCED) in 1992, was the first International Environmental Convention to develop measures for the use and protection of TK, related to the conservation and sustainable use of biodiversity. ***CBD:* Article 8 (j)** encourages National Governments to develop mechanisms for equitable benefit sharing.

5.3.2. National Initiatives

NATIONAL INITIATIVES

Protection of TK

Defensive

- National Biodiversity Authority
- TKDL project of Information technology Division of CSIR
- Capable of protecting 2 lakhs medical formulations similar to Neem and Turmeric
- Signed an agreement with European patent office (EPO)

Positive

- The Patents (Amendment) act, 2005
- Biodiversity Act, 2002, Rules 2004
- Systematic documentation, scientific validation of TK, process/product development, technology transfer, commercialization and benefit sharing. Implementation of article 8(j) of CBD
- Scheduled Tribes and other Traditional Forest Dwellers (Forest Rights) Act, 2006
- To develop viable models under *sui generis* system

TBGRI INITIATIVES

Protection of TK

Defensive

- Systematic documentation of TK from the oral tradition
- Prior informed consent (Appendix)
- Documentation of disclosed/partially disclosed/undisclosed information
- Digitalization of TK based on selected disclosed information
- Database station
- Access to data by the end users by signing an agreement

Positive

- Viable models developed under *sui generis* system
- Establishment of ownership rights?
- Formulation of traditional knowledge holders Rights/Act?
- Scientific validation based on TK process/product development, technology transfer, commercialization and benefit sharing.

5.4. TBGRI Model for Benefit Sharing

(a) ABS Kani Model Of Access And Benefit Sharing

The issues of Access and Benefit Sharing (ABS) under the CBD has been widely discussed and debated at global level but it still remains a technically challenging and legally complex issue. Jawaharlal Nehru Tropical Botanic Garden and Research Institute (JNTBGRI) Palode, Thiruvananthapuram during the year 1996 developed a unique benefit sharing model, popularly known as **TBGRI model of benefit sharing**. for a scientifically validated herbal drug, **'Jeevani'** based on a tribal lead. The major players involved in the transaction are TBGRI and the Kani tribe. Thus, for the first time in India and perhaps in the world, the tribal intellectual property rights were recognized, in a dignified manner. This unique benefit sharing model, popularly known as TBGRI model for benefit sharing, won the **'UN-Equator Initiative Award'** in 2002. The Kani case has offered lessons to be learnt at various levels. "It also illustrates the growing interplay between the collective rights, as evident from the common goods approach and individual rights, as evident from the upcoming intellectual property regime. With the expansion of economic growth across developing countries, it is this common goods approach which is facing the maximum pressure from those who wish to tap natural resources for personal profits [7]".

(b) Contractual Agreement (Mutually Agreed Terms)

This is an agreement signed with Shri T.M Shahul Hameed Vaidyar, a traditional healer from Kollam district of Kerala for conducting pre-clinical study of a poly herbal formulation based on traditional knowledge with multiple therapeutic effects as anti-diabetic, anti-fatigue, hepato-protective and anti-oxidant. It is noted that, he has been included (Traditional Healer – Non tribe) as an inventor in the patent application (Patent filed on 04/07/2011, application No. 2277/CHE/2011.

Appendix

Model of Prior Informed Consent Form

Developed by TBGRI Exclusively for the Systematic Documentation of Traditional Knowledge Related to Plants Used for Food and Medicine

I. Has it been explained to you, that you have complete discretion on whether to disclose or not, the TK you hold?

Yes		No	

II. Were the advantages and disadvantages of disclosure of TK made known to you?

Yes		No	

III. Were you explained in brief the Biological Diversity Act and your rights under the Act?

Yes		No	

IV. Is the knowledge disclosed by you exclusively known to you?

Yes		No	

If No, who else knows about the Disclosed Knowledge?

All members of the Community

Traditional healers of your community

Locally known

If known only by the Traditional Holders of the Community, a brief explanatory note:

V. * Whether you approve that the knowledge disclosed by you is published in a scientific journal or other journal or any other media with the intention of inviting enquiries for commercialization/research?

Yes	No

VI. * Whether you approved this knowledge to be protected by the existing Intellectual Property Rights?

Yes	No

VII. * Whether you approve of research/commercialization of the knowledge with improvements?

VIII. * What kind of benefit sharing is preferred?

Yes	No

a. Monetary b. Non-Monetary

If, Non – Monetary, what kinds of Benefit Sharing are you interested in?

I/WE have been explained every question in a language known to me/us. All answers are given by me/us after understanding the questions completely and the implications thereof. All information given are true to the best of my/our knowledge and belief.

Signature:

Place: Name:

Date: Contact Address:

* Answer to these shall depend on the answer to Q. NO.4 in case, the knowledge belongs to the Community or Traditional Healers, due care shall be taken to check who is authorized to give the consent.

ACKNOWLEDGEMENTS

Declared none.

CONFLICT OF INTEREST

The author confirms that this chapter contents have no conflict of interest.

REFERENCES

[1] Udayakumar Gali, Prasanna Kumar WG, Suneeta M. Biodiversity (Resource Guide for teachers and students). Andhra Pradesh National Green Crops, Hyderabad 2012; pp. 10.

[2] Rasjasekharan S, Latha PG, Mohanan N, Mathew Day, Vinodkumar TG Nair, Navas M, Vimalkumar CS. Students Handbook on Medicinal and Food Plants. JNTBGRI, Thiruvananthapuram 2013; pp. 124.

[3] Padmanabha Pillai Sastri. 'Vrikshayurveda' of Sarangadhara Acharya. The National Printing Works, Trivandrum 1946; pp. 1-65.

[4] Rajasekharan S, Mohanan N, Latha PG, Mohanlal KG, Nair GM. Stars and Trees. TEPS and JNTBGRI, Thiruvananthapuram 2005; pp. 1-100.

[5] Rajasekharan S, Venkada Subramanian N, Bijukumar A. Report on Development of Gene pool Conservation Area. Western Ghat Development Programme, Kollam District of Kerala 2005; pp. 3-4.

[6] Rajasekharan S. Traditional and Folk Practices-Contemporary Relevance and Future Prospects. Proceedings of Kerala Environment Congress, Thiruvananthapuram 2013; pp. 141-156.

[7] Rajasekharan S, Puspangadan P, Latha PG, Vinodkumar TG Nair. Learning from Traditional Knowledge Intellectual Property, Traditional Knowledge; Community Perspective on Access and Benefit Sharing-ABS Kani Model of Access and Benefit Sharing. JNTBGRI & KSCSTE, Thiruvananthapuram 2012; pp. 1-27.

Chapter 5: Current and Likely Trends in Biodiversity Conservation

CHAPTER 5

Current and Likely Trends in Biodiversity Conservation

Rajmohan K.*

Kerala Agricultural University, College of Agriculture, Vellayani, Thiruvananthapuram, Kerala, India

Abstract: Crop and livestock genetic resource conservation must be strengthened. Botanic gardens, zoological parks and public aquaria must be strengthened. Human Capacity to Conserve Biodiversity must be improved. Global biodiversity information network has to be established. Basic and applied research on biodiversity conservation must be promoted. Increased public participation, respect for basic human rights and improved access to education and information are essential elements of biodiversity conservation. The costs and benefits of biodiversity conservation should be shared equitably among nations and people.

Keywords: Alpha diversity, Beta diversity, biodiversity, botanic gardens, cDNA library, conservation, cryopreservation, Gamma diversity, gene bank, habitat loss, sacred forest, tissue culture.

1. INTRODUCTION

Biodiversity is the variability within and between ecosystems, species and genes [1]. It is the biological wealth of the world evolved as the product of events during millions of years. Genetic diversity is the variation of genes within species, denoting distinct populations of the same species as well as genetic variation within a population. Genetic diversity helps a population to adapt to its environment. It will enable the population to respond to natural selection. Species diversity represents the different species within a region. Alpha diversity, Beta diversity and Gamma diversity represent the diversity at community and ecosystem levels. Alpha diversity refers to the diversity of organisms sharing the same community. Beta diversity represents the rate of replacement of species along a gradient of habitat. Gamma diversity is the rate at which additional species appear in different localities of the same habitat, as replacement.

***Corresponding author Rajmohan K.:** Kerala Agricultural University, College of Agriculture, Vellayani, Thiruvananthapuram, Kerala, India; Tel: +91 9526467549; +91 471 2456881; Fax: +91 471 2382239; E-mail: rajmohan33@yahoo.com

Laladhas K.P., Oommen V. Oommen and Sudhakaran P.R. (Eds.)

Biodiversity is also enriched by its expressions. These are represented by the comparative abundance of species, age groups of an existing population, regional structure of communities and their periodic changes *etc*. Predation, parasitism, and mutualism are also important in this respect. Biodiversity includes agricultural biodiversity, dry land biodiversity, forest biodiversity, inland water biodiversity, marine biodiversity and mountain biodiversity, as well. Even, human cultural diversity can be considered as part of biodiversity. Hot spots of biodiversity represents the richest and most threatened reservoirs of plant and animal life. India ranks twelfth among the leading biodiversity centres of the world [2].

2. THREATS TO BIODIVERSITY

It has been reported that the present species extinction rates are about 1000 folds of that in pre-human times. This is quite alarming and highly detrimental to a sustainable society. This drastic loss of biodiversity, involving the erosion of genes, species, and ecosystem is due to several factors [3]. Increasing population growth, high rate of resources consumption, fast rate of urbanization and rising pollution are among them. It has been estimated that about sixty per cent of ecosystem services worldwide have become degraded. This has happened mostly during the recent past. The primary contributing factors for this degradation have been pointed out as land- and ocean-use practices. It is of great concern to note that we are losing biodiversity even before its full development. This in turn affects the evolutionary capability of biota in the context of environmental changes. This can very well lead to the extinction of species [4].

The loss of species at very slow rate is designated as natural or background extinction. This happens with changes in environmental conditions while some species are replaced by new ones that are better adapted to altered conditions. Mass extinction is a major cause of biodiversity loss. It is the process of extinction of large number of species due to natural calamities. Biodiversity loss is also happening through the introduction of exotic species. Even Agriculture leads to the extinction of traditional varieties and cultivars. Agriculture causes the reduction of crop genetic base, as it promotes monoculture. This in turn results in increased vulnerability of crops to pests and diseases. Human interventions have accelerated the natural extinction rate of species [4]. Human interventions like hunting, chemical control of pests and diseases, introduction of exotic species and intensive cropping leads to habitat loss pollution and deforestation. Together with climate change, this situation can result, in huge loss of biodiversity of plants, animals and micro organisms in the near future [5]. It is the need of the hour to safeguard biodiversity, by way of retaining the structure and functions of the natural systems that support biodiversity.

3. BIODIVERSITY CONSERVATION

Conservation of biodiversity is of paramount importance. It is essentially the natural human life support system. Optimum utilization of biodiversity, maintaining it and sharing its benefits to the entire humanity is a way of living in harmony with nature. While facilitating development of the society, it will protect the biotic wealth. It is a misconception that biodiversity conservation targets the protection of wild species only. It should protect the genetic diversity of cultivated and domesticated species as well.

It is essential to understand the factors contributing to the current loss of biodiversity. It is also important to understand the enormous opportunities that genes, species, and ecosystems provide for sustainable development. These are important in evolving effective biodiversity conservation strategies. Urgent steps should be taken to prevent further loss of biodiversity, by way of species extinction, loss of habitats and genetic diversity erosion. All these are irreversible. We should aim at the sustainable utilization of biodiversity, saving it and sharing the benefits to all. Our biodiversity conservation strategies, while aiming at the prevention of the loss of biodiversity should protect the populations, habitats and ecological cycles supporting biodiversity. Conventional biodiversity conservation programmes involve *in situ* and *ex situ* methods. *In situ* conservation involves the establishment of biosphere reserves, national parks and wild life sanctuaries and other protected areas. It essentially helps to maintain gene banks.

Biosphere reserves preserve genetic diversity in representative ecosystems. They are multipurpose protected areas. They are helpful in protecting wild populations. Also, they support the traditional life style of tribals and aborigins. Protection of domesticated plant and animal genetic resources is also aimed at. Biosphere reserves are divided into core, buffer and manipulation zones. In core zone no human activity is allowed. Limited human activity is permitted in the buffer zone. Several human activities can occur in the manipulation zone. National park essentially is an area strictly reserved for the betterment of the wildlife. Here activities like forestry, grazing or cultivation are not permitted. Wild life sanctuary is a protected area, reserved for the conservation of animals. Sacred forests are forest patches protected by communities due to religious belief, without any human interventions. These are characterized by the presence of a number of rare, endangered and endemic species.

In the process of extinction, the population of a species may become fragile and fragmented. It may finally become extremely difficult for it to survive in its natural habitat. Also, its conservation in the wild in its normal natural state may not be possible. *Ex-situ* or off site conservation becomes relevant in this context. Representative samples of the species are propagated and conserved *ex situ*, in botanic gardens, zoological gardens, aquaria and research centres, where they can be subjected to studies and exhibited to the public. *Ex situ* conservation has been observed to facilitate the recovery of species, associated with increased propagation efficiency. It is also possible to re-introduce the species in their natural habitat, after some time.

Ex situ conservation adopts the establishment of gene banks in the field as well as laboratories. Gene banks facilitate the availability of crop varieties and land races for the use of farmers and research workers. In *vitro* conservation, including cryopreservation is an important method adopted for *ex-situ* conservation of plants. Cryopreservation of plants will help in preserving species and varieties, especially endangered ones. Preserving mammalian sperms will facilitate *in vitro* fertilization and surrogacy for saving species. Establishment of zoo is an important practice for the *ex situ* conservation of animals.

Conventional methods for the conservation of biodiversity should be supplemented and supported by modern techniques. Understanding biodiversity and biological resources with the help of modern technology is important. Biotechnological tools can be effectively utilized for the conservation of biodiversity. Tissue culture, cryopreservation, DNA banking, Bar coding, microbial repository *etc.* are relevant in this respect. Future of conservation biology depends to a large extent on biotechnological tools. Genomics can be effectively utilized for the conservation and management of genetic resources. This will help in understanding their genetic base and facilitating suitable genetic modifications for better adaptation and utilization. cDNA libraries are useful for the conservation of biodiversity. There are several genetic techniques available for assessing the genome of a species, understand population structure and design specific conservation strategies. Molecular markers can effectively be used for analyzing the differences between individuals and populations. They include Random Fragment Length Polymorphisms, Amplified Fragment Length Polymorphisms, Random Amplified Polymorphic DNA, minisatellites, microsatellites, Single-nucleotide polymorphisms *etc.* They can provide information useful for the long-term conservation of genetic diversity. DNA barcoding is a modern tool useful to understand the diversity of species within an ecosystem, and evaluate the genetic variability within species. Another powerful modern tool adopted for biodiversity conservation, in relation to the problem of deforestation is remote-sensing technology. Geographic Information Systems can analyse landscape ecology and adopt spatial analysis for addressing these issues.

Efficient strategy for biodiversity conservation should aim at the study and saving of biodiversity, while ensuring the sustainable and equitable use of its benefits. Documentation of the composition, distribution, structure, and function of biodiversity is important. It is also essential to understand the roles and functions of genes, species, and ecosystems. Along with the threatening factors, it should consider the opportunities that genes, species, and ecosystems provide for sustainable development. Maintenance of natural habitats and eco systems is an indirect way of maintaining species and gene pools. Effective steps should be taken for preventing the degradation of key natural ecosystems. Land that have been disturbed should get special attention for maintaining diversity. Better understanding of biodiversity and its role in ecosystems and human life will facilitate to develop effective strategies for slowing down the loss of biodiversity.

The UN General Assembly declared the year 2010 as the International Year of Biodiversity. The Conference of the Parties in Nagoya meeting in 2010, adopted new biodiversity targets. As per this the direct drivers (climate change, resource consumption, pollution, marine, freshwater and terrestrial habitat fragmentation and destruction, unsustainable use of wild living resources, alien invasive species, emerging infectious diseases) and indirect drivers (resource consumption, globalization, trade, demographic drivers) need to be addressed specifically and explicitly. In order to address the indirect drivers of biodiversity loss public and private educational curricula should include the value of biodiversity and the measures to conserve it. Traditional knowledge and the rights of indigenous people have to be protected. Present scientific information on biodiversity, has to be improved and shared with decision makers. At the national level reforms of the existing public policies that controls the waste or misuse of biodiversity need to be made. Forestry policies that encourage resource degradation and the conversion of forest ecosystems must be abandoned. Policies that cause degradation and loss of biodiversity in coastal and marine ecosystems have to be changed. Agricultural policies that promote excessive uniformity of crops and crop varieties, or the overuse of chemical fertilizers and pesticides must be eliminated. Recycling and conservation of resources should be promoted. Biodiversity conservation initiatives in the private sector have to be supported. Actions to conserve biodiversity through international cooperation should be promoted. Mechanisms involving scientists, non-governmental organizations, and policy-makers to provide guidance on the protection, understanding, sustainable and equitable use of biodiversity must be established. Early Warning Network, to monitor potential threats to biodiversity must be established. Biodiversity conservation must be integrated into international economic policy.

ACKNOWLEDGEMENTS

Declared none.

CONFLICT OF INTEREST

The author(s) confirm that this chapter contents have no conflict of interest.

REFERENCES

[1] Larsson, T. B. 2001. *Biodiversity Evaluation Tools for European Forests*. Wiley-Blackwell. p. 178.

[2] Khoshoo, T. N. 1995. Census of India's biodiversity: tasks ahead. *Curr. Sci.*69: 14-17.

[3] Pimm, S.L., C.N. Jenkins, R. Abell, T.M. Brooks, J.L. Gittleman, L.N. Joppa, P.H. Raven, C.M. Roberts and J.O. Sexton. 2014. The biodiversity of species and their rates of extinction, distribution, and protection. *Science* 344(6187): 1246752

[4] Pimm, S. L., G. J. Russell, J. L. Gittleman, and T. M. Brooks. 1995. The future of biodiversity. *Science* 269 (5222): 347–350.

[5] Longcore, T. and C. Rich. 2004. Ecological light pollution. *Front Ecol Environ* 2(4): 191–198.

CHAPTER 6

Natural Bioresources of North Eastern Region of India - Its Conservation and Proper Utilization

Unni B.G., Bharadwaz M., Kalita M., Dey T., Gogoi K., Kalita J., Islam S.J. and Dutta P.

Biotechnology Division CSIR-North East Institute of Science and Technology, Jorhat- 785006, Assam, India

Abstract: Natural bioresources refers to all kinds of naturally occurring resources of biological origin. Biodiversity or biological diversity refers to the diversity of any kind of living organisms that exists in that particular area, region and country or on the earth. The North Eastern region of India comprises of the states of Arunachal Pradesh, Sikkim, Mizoram, Assam, Manipur, Nagaland, Meghalaya, Tripura. This region of India is rich in biodiversity due to high rainfall and plenty of sunlight. This region can be addressed as the geographical 'gateway' for endemic flora as well as fauna. This area is one of the richest biomes of the world, endowed with a wide range of ecological conditions ranging from low lying areas of Assam to Kanchenzanga of Sikkim. The North East India represents a wide range of forests ranging from tropical to sub – tropical, temperate to alpine regions. Forests of this region consist of a huge number of endemic plants and animals. Two UNESCO World heritage sites namely, Kaziranga and Manas are also located in the North East belt (Report UNEP World Conservation Monitoring Centre, UK). However, the valuable resources of this part of the country is depleting at an alarming rate due to unavoidable human activities, colonization, urbanization, agriculture, mining, transport, and technology. The natural hazards like floods, landslides are also adding to the loss of biodiversity. The conservation of the resources and their proper utilization with the modern techniques have long been realized in the region and it still requires scientific intervention and concerted efforts to ensure sustainable use of these resources and livelihood security.

Keywords: Antibacterial activity, antimicrobial activity, biodiversity, bioresources, botanical garden, deforestation, endangered, endemism, fauna, flora, hotspots, medicinal plants, microorganisms, natural calamities, North East India, poaching, pollution, rainforest, sericulture, silkworm, sustainable utilization.

1. INTRODUCTION

The planet earth is inhabited by several millions of species under about 100

***Corresponding author Unni B.G.:** Biotechnology Division CSIR-North East Institute of Science and Technology, Jorhat- 785006, Assam, India; Tel: 0376-2372951; Fax: 0376-2370011;
E-mail: bgunni@yahoo.com

Laladhas K.P., Oommen V. Oommen and Sudhakaran P.R. (Eds.)

different phyla, but many species are yet to be discovered [1]. Every year above 15,000 new species are described and many new species are still evolving during our lifetime. In the meanwhile depletion of existing species is occurring at an alarming rate mostly due to habitat alteration caused by human activities. "India with an area of 3,287,263 square km is Asia's second largest nation and the seventh largest country in the world [2]" and occupies 10th position among the plant rich nations and accounts for 10.78 % of the global flora [3].

The North East India on account of its varied physiography, climate and soil conditions harbours a rich flora and is one of the most biodiversity rich hotspot under Indo Burma megabiodiversity [4, 5]. The tropical vegetation in general occurs up to an elevation of 900 m in most of the parts of North-East India and 27% of total area are under forest cover [6]. This region is represented by more than 500 flowering plant species and constitutes 30-40% of the total flora of the country [6]. "The forest types occurring in the region are evergreen (where rainfall exceeds 2300 mm per annum) and semi-evergreen rain forests, moist deciduous monsoon forests, riparian forests, swamps and grasslands [7]". Evergreen rain forests are found in the Assam Valley, eastern Himalayas and the lower parts of the Naga Hills, Meghalaya, Mizoram, and Manipur. The NE region extends from foothills of Himalayas in eastern range [8] and has been the centre of attention to biologists & naturalists for its high biodiversity [9]. The forests of this region are said to be the true tropical rainforest.

2. BIODIVERSITY OF NORTH EASTERN REGION

"Out of 15,000 species of flowering plants in India, more than 8,000 species are found in the North Eastern region, which includes 40 species of gymnosperms, 500 species of pteridophytes, 825 species of orchids, 80 species of rhododendrons, 60 species of bamboo and 25 species of canes in 51 different forests [8]". "These forests include tropical moist deciduous forests, tropical semi evergreen forests, tropical wet evergreen forests, subtropical forests, temperate forests and alpine forests [8]". "North Eastern region of India can be geographically divided into three types, *viz.* (i) Meghalaya plateau (ii) North-eastern hills and basin and (iii) the Brahmaputra valley [10, 11]".

Most of the areas of North-East India have remained unexplored for medicinal plants, microbes, silkworms and endemic plants and animals. The demand for medicinal plants is increasing day by day, and can serve as a potential resource of economic development [12]. There are around 350 species of medicinal plants recorded in Assam and 450 species in Arunachal Pradesh. In Meghalaya, 377 species of medicinal plants are used by locals for primary health care. 375 species of plants having medicinal value have been recorded in Manipur, more than 100 species in Nagaland, around 424 species in Sikkim and more than 210 species in Tripura. This part of the country is inhabited by large number of ethnic or tribal people, who possess a vast traditional knowledge for effective use these medicinal plants as herbal medicines.

Wild silk-moths are an indispensable part of the biodiversity of North East India. It includes - Muga silk worm (*Antharaea assamensis*), Eri silk worm (*Philosamia ricini*), Oak Tasar silk worm (*Antheraea proyeli*), and Mulberry (Pat) silk worm (*Bombyx mori*). Except the silks produced by *Bombyx mori*, other varieties of silks are generally termed as non-mulberry silks [13]. Large quantity of cocoons of different silkworms is utilized to produce various types of textiles. India, especially the state of Assam has the unique distinction of producing all types of silks commercially. Sericulture is also a common labour intensive industry in Arunachal Pradesh. The state of Arunachal Pradesh produced around 12.38 Metric tons of raw silk during 2007–2008. However, the production has reduced considerably in the recent years. In Meghalaya, four different types of Eri silk worms are found to be reared in Khasi-Jaintia and Ri-Bhoi hills by farmers. Manipur was also once a key staging post on the fabled silk route. However, in the recent times Manipur accounts for just a fraction of the total silk produced by India. For centuries, the women of Khurkhul village in Manipur have raised silkworms. Silkworm caterpillars are also a common sight in Nagaland. People not only produce silk in this place but also consider it to be a delicious food item. Sikkim is also rich in arthropods, especially butterflies and silkworms. Wild silkworms are generally observed in natural colours – cream, beige, brown and gold. However, the silk moths of North East India have decreased surprisingly owing to drastic climate change, deforestation and uncontrolled pests.

The North East India is one among the 34 Hotspots of the world. There are seventeen endemic birds found in Assam, and these birds could be seen in grassland, bushes, bamboo thickets, dense evergreen forest, and in high grassland on riverbanks. The Asian elephants and the one horned rhinoceros are the pride of Assam. The Golden Langur, a species described in the 1960s, is endemic to a small region between the Manas and Sankosh Rivers, two tributaries of river Brahmaputra in the state of Assam. Hoolock Gibbon, the only ape in India is found in Assam, Manipur and Tripura. Assam is also the homeland of several species of orchids. It is also home to the endangered White-winged Wood Duck (*Asarcornis scutulata*). The bioresource rich place of Arunachal Pradesh has many rare, endangered and endemic plants, 25 species of mammals, 7 species of primates, with 4500 species of Angiosperms including 550 species of orchids, 73 species of bamboos, 18 species of canes and 34 species of Gymnosperms. Meghalaya is famous for its endemic Pitcher Plant. Manipur has over 75 species of endemic plants and is the home of the dancing deer, Sangai. It has 4 species of primates and around 5 species of carnivores. Approximately 5000 species of flowering plants, 11 oak varieties, 515 rare orchids, 36 rhododenderon species, 60 primula species, 23 bamboo varieties, 16 conifer species, 362 types of ferns are endemic to Sikkim. Tripura has over 456 endemic plant species. Spectacle monkeys are endemic to Tripura. Tragopan one of the most beautiful birds among pheasants is endemic to Nagaland. The eco region of Mizoram is the only habitat to a species of bat and a rodent and around 6 species of birds.

The medicinal plants, silk moths, endemic flora and fauna and many other species of this region are now in the list of endangered species due to various human and natural activities like;

1. Alteration of habitats, absence of cover or shelter.
2. Deforestation.
3. Hunting and poaching for food, fur, musk, tusk and horn.
4. Pollution.
5. Various natural calamities.

If the natural resources are not maintained, the balance of nature will be disturbed. Natural resources help to study morphology, anatomy, physiology and ecology of different species [14]. The natural resources are the only assets of North East India. Therefore it is must be mandatory principle of every citizen to conserve and utilize the natural resources with proper management [15].

2.1. Endemic Animals and Birds of North-East India

Many animal species are found in Manipur. Clouded Leopard and the Golden Cat are the rare, carnivorous nocturnal species among them. "The Malayan Sun Bear (*Helarctos malyanus*) is rare and elusive species restricted to the Yangoupokpi Lokchao Wildlife Sanctuary of Manipur [16]". The Sangai (*Cervus eldi eldi*) is the endemic species, which feeds on the floating mass of vegetation present in Loktak Lake of Manipur.

Arunachal Pradesh is home of many endemic species of animals and birds. It is perhaps the only State having the four major cats (tiger, leopard, clouded leopard and snow leopard), seven species of primates (hoolock gibbon, slow loris, Assamese macaque, rhesus macaque, pig-tailed macaque, stump-tailed macaque, and capped langur) and three goat antelopes (serow, goral and takin). The highly endangered hispid hare (*Caprolagus hispidus*) is also found in low grassy areas of the State. Three endangered species of vultures- white-rumped vulture (*Gyps bengalensis*), Indian vulture (*Gyps indicus)* and slender-billed vulture (*Gyps tenuirostris)* are found at the Dying Ering Wildlife Sanctuary in East Siang district of Arunchal Pradesh. At present the White billed Heron is only found in Namdapha National Park. According to Zoological Survey of India, a rare species of bird, white winged wood duck (*Asarcornis scutulata*) has been recorded in Namdapha National Park. Spectacled langur is an endemic species fauna in Sepahijala wildlife sanctuary of Tripura. The crab eating Mongoose is also found in this sanctuary. "Sunda Slow Loris, Phayre's Leaf Monkey, Hoolock Gibbon, Leopard Cat, Marbled Cat, Golden cat, common otter, and Indian elephant are abundant in Tripura [9]".

"The snow leopard, Himalayan tahr, musk deer, red panda, Himalayan serow, Himalayan goral, Himalayan marmot, muntjac, asian black bear, common langur, clouded leopard, leopard cat, dhole, marbled cat, tibetan wolf, hog badger, Himalayan jungle cat, binturong are the important fauna of Sikkim. A few endangered species like musk deer, red panda, Himalayan black bear, serow (long-haired goat) and some species of pheasants are also found in the Kyongnosla Alpine Sanctuary [17]".

Out of five species of Tragopan, one of the most beautiful birds found in the Himalayan region, *Tragopan blythii* is found only in Nagaland. "Secondary dogs, porcupines, elephants, pangolins, leopards, many species of monkeys, sambar, bears, harts, oxen, and buffaloes are found in the state. Mithun is found only in this state among the North Eastern states of India [18]".

Assam, a biological hotspot is home to rare and endemic animal species. The Indian one horned Rhinoceros (*Rhinoceros unicornis*), Spotted Deer, Tiger, Swamp Deer, Hoolock Gibbon, Clouded Leopard, Pigmy Hog, Golden Langur (*Trachypithecus geei*), Hispid Hare, Golden Cat, Giant Civet, Binturong, Porcupines, Hog Badgers, Civet cats *etc*. are found in Assam. Besides, Gangetic Dolphins, Mongoose, Giant Squirrels and Pythons are found here. "Blue-throated Barbet, Ring-tailed Fishing Eagle, the Himalayan Golden-backed Three- toed

Wood-pecker, Tawny-breasted Wren-babbler, Blackish-breasted Babbler, Snowy-throated Babbler, Striped Laughing thrush, Grey Sibia *etc.* are some endemic birds of Assam [18]".

Mizoram is the home state of several endemic animal species. About 149 mammalian species are found here, including two near-endemic species a bat (*Pipistrellus joffrei*) and a murid rodent (*Hadromys humei*). "It is also a home to several endangered species like the Tiger, Asian Elephant, Himalayan Goral, Smooth-coated Otter, Indian Civet, Red Panda, Clouded Leopard, Back-Striped Weasel, Eld's Deer Gaur, *etc.* [18]".

The important mammal species found in Meghalaya are elephants, civets, weasels, mongooses, bear, rodents, wild buffalo, deer, wild boar and a number of primates. Balpakram National Park is home of the Red Panda and Selbagre Hoolock Gibbon Reserve is home to the Hoolock Gibbons. Prominent bird species found in Meghalaya are Magpie-Robin, Hill Myna, Large Pied Hornbill, Red-vented Bulbul and Great Indian Hornbill, Large Indian Parakeet, Peacock Pheasant, Common Green Pigeon and the Blue Jay *etc.*

3. NATURAL BIORESOURCES OF NORTH EASTERN REGION

3.1. Medicinal Plants

The present work is the result of intensive study of almost all the medicinal plants of the north-east region. A comparative study of antibacterial property of some of the medicinal plants of NE region was carried out against *Pseudomonas aeruginosa* strain AC-3 causing flacherie in muga silkworm, *Antheraea assamensis*. A fruit based bioformulation of *Terminalia chebula* (Fig. **1**), can be effectively applied for reducing the intensity of flacherie disease in *Persea bombycina*, the principal food plant of *A. assamensis*; *R. communis* and *S. ricini* and for enhanced cocoon production with uniform shape size and improved quality of silk.

Few of the medicinal plants used for health care and its habitat are explained here. However, large numbers of such plants are yet to be identified and studied in detail for their ethnobotanical aspects. The common medicinal plants available in North Eastern region of India along with their uses are presented in Table **1** [4, 19-24].

Figure 1: (a) Raw and (b) dried fruits of *T. chebula.*

Table 1: Medicinal plant used for health care and its habitat within North Eastern states

Vernacular/ Common Name	Species	Parts of Plant Used	Use
Agar	*Aquilaria agalocha* Roxb	Bark, root & heartwood	Inflammation, arthritis, vomiting, cardiac disorders, asthma, leprosy, headache, gout
Kasa	*Saccharum spontaneum* L.	Roots	Dyspepsia, burning sensation, piles and sexual weakness
Bhurja	*Betula utilis* D. Don	Stem, bark	Tridosa, convulsions, leprosy, diarrhea, epilepsy,bronchitis jaundice
Dugdhika	*Euphorbia thymifolia* L.	Whole plant	Skin disease-Allergy & itching, skin disease-ringworm
Khamalu	*Dioscorea floribunda* M.	Tuber	Used as a vitalizer
Narang	*Citrus medica* L.	Fruits	Maintain hydration, treatment of tuberculosis, stomach upsets, cold, cough & respiratory disorder
Sarpagandha	*Rauvolfia serpentine* L.	Root	Diarrhea & dysentery, scabies, snake bite & scorpion sting, stomach disorders
Amrul	*Oxalis corniculata* L.	Whole plant	Eye disease, hair & scalp problems, jaundice & related problems, scurvy, eczema, worm infection
Xilikha	*Terminalia chebula* Retz.	Fruit	Diarrhea & dysentery, fever & malaria, respiratory disease, worm infection
Indian plum	*Ziziphus mauritiana* Lam.	Bark, Fruit	Cancerous disease, epilepsy, throat problem
Dhataki	*Woodfordia fruticosa* (L.) Kurz	Flower, Bark	Piles, liver problems, gastric troubles
Rajbrikha	*Cassia fistula* L.	Fruits, Leaves	Jaundice & related problems, leprosy,

Table 1: contd…

			ringworm, stomach disorders
Tong-loti, Bonmala	*Callicarpa macrophylla* Vahl.	Bark	Cuts & wounds, gynecological disorders
Bleeding heart	*Clerodendrum infortunatum* L.	Leaves	Hypertension, diarrhea, & dysentery, fever & malaria
Dhopat tita	*Clerodendron viscoum Vent.*	Bark /leaves	Allergy & itching, renovating & sexual debility
Common Fox glove	*Digitalis purpurea* L.	Leaves	Congestive heart failure, edema, asthma, epilepsy, tuberculosis, constipation
Kukurtarul	*Dioscorea deltoidea* Wall. ex Kunth	Barks and tubers	Good source of Steroidal hormone
Yam	*Dioscorea composita* L.	Tubers	Joint pain, diabetes, oral contraceptive
Somlata	*Ephedra gerardiana* Wallich	Whole plant	Asthma, hay fever & allergic complaints
Kurkure jhar	*Equisetum debile* Roxb.	Aerial part	Rheumatism, stomach, infection, inflammation
Rudraksh	*Elaeocarpus ganitrus* Roxb.	Seed	Stress, anxiety, depression, lack of concentration, prolonged cough
Nageswari	*Mesua ferrea* L.	Bark	Epilepsy
Sajana	*Moringa oleifera* L.	Root, bark and fruits	Allergy & itching, eczema, ringworm
Semal	*Bombax cebia* L.	Exude gum	Gynecological disorders, piles, urinary problem
Arka, Akon	*Calotropis gigantean* (L.)	Whole plant/ root & bark	Respiratory disease, rheumatism & lumnbago, leucoderma, leprosy, ringworm
Guggul	*Commiphora wightii* (Arn.)	Resin	Rheumatic & anti inflammatory infection, chronic ulcers
Misimi tita, Nanura	*Coptis teeta* Wall	Whole plant/Root	Skin disease, sore throat
Rhododendron	*Rhododendron arboretum* Smith.	Leaves, flower, roots and bark	Gout, rheumatism, diarrhea, blood dysentery
Cinchona bark	*Cinchona rubra* L.	Bark	Malaria, regulates heart beat, stimulates digestion
Aksoda, Akhrot	*Juglans regia* L.	Bark, fruit and seeds	Diarrhea, arthritis, asthma, eczema, Helminthiasis, sinusitis, diabetes mellitus
Tagara	*Valeriana jatamansi* Jones	Rhizome and roots	Hypertension, gastro intestinal problems, epilepsy, cough, lung congestion
Kutki	*Picrorhiza kurroa* Royle	Rhizome	Liver disease, Jaundice
Musta	*Cyperus rotundus* L.)	Root and tubers	Stomach & bowl disorders, inflammatory diseases, dysmenorrheal& menstrual

Table 1: contd…

			irregularities
Greater galangal/ Rasna	*Alpinia galanga* (L.) Wild.	Rhizome	Cold, cough & bronchitis
Snuhi	*Euphorbia nerifolia* L.	Roots and plant latex	Abdominal troubles, bronchitis, tumors, leucoderma, piles, ulcers
Lakuca	*Artocarpus lakoocha* Roxb.	Root	Skin ailments, headache
Pitalai	*Stereospermum chelonoides* (L.f.) DC.	Roots, barks and flowers	Fever & malaria
Phalgu	*Ficus hispida* L.f.	Roots	Skin disease-Ringworm
Jamun	*Syzygimum cumini* (L.)	Fruit	Diabetes, diarrhea & dysentery, throat problem
Ghora-neem	*Melia azedarach* L.	Leaves	Leprosy, anthelmintic, diuretic, burns, chicken pox, small pox, piles
Bitter vine	*Mikania micrantha* Kunth.	Whole plant	Respiratory disease, tuberculosis
Bhatgeela	*Oroxylum indicum* (L.)	Seeds and bark	Jaundice, Scabies, Stomach disorders, cold, cough & bronchitis, fever, malaria
Wild bhindi	*Abelmoschus moschatus* Medik.	Seeds	Gynecological disorders, respiratory disease
Fischer monkshood	*Aconitum fischeri Reich.*	Roots	Cough & fever, fatigue, lumbago, small pox, ulcers, abortions
Kyohan	*Zanthoxylum acanthopodium*	Leaf and roots,	Toothache, common cold, cough, fever
Amra	*Spondius magnifera* Wild.	Leaf extract	Dental problem, anti-inflammatory, urinary problem, skin problem
Hatush	*Morus australis* Poir	Milky juice	Nasal bleeding, skin disease, eczema, scabies
Bellthi	*Passiflora edulis* Sim.	Tender leaves	Blood pressure, diarrhea and dysentery, jaundice.
Lajuki lota	*Mimosa pudica* L.	Leaves	Dental problem, Cancerous diseases, Gynecological disorders, skin diseases, Urinary Problems, Rheumatism
Mhejung	*Adhatoda vasica* Ness.	Leaves	Asthma, cough, anti-ulcer activity, bronchitis, tuberculosis
Khokshu	*Cassia hirsute* L.	Dried seeds	Stomach disorder
Himalayan Yew	*Taxus baccata* L.	Whole plant	Ovarian cancer, breast cancer, Asthma, bronchitis, rheumatism, epilepsy
Tejphal	*Zanthoxylum*	Leaves and fruits	Measles, skin disease, mental disorder

Table 1: contd…

	armatum DC.		
Sako	*Aacia pennata* Linn.	Whole plant	Dysentery and urinary discharges, anti-inflammatory, spasmolytic
Ruja	*Diospyros lanceaefolia* Roxb.	Tree, bark and fruit	Skin disease, fruits are powered & sprayed in streams to poison fish
Reetha	*Sapindius mukorossi* Gaertn.	Tree and fruit	Eczema, psoriasis, removing freckles
Vedailata	*Paedaria foedita* Linn.	Leaves, roots	Arthritis & rheumatic disorders
Era	*Ricinus communis* Linn.	Leaves	Abortion, Burn, Dental problem, Headache, Measles, allergy itching, Snake bite
Manimuni	*Centella asiatica* Linn.	Whole plant	Blood pressure, Jaundice, Skin disease, Stomach disorders
Nephaphu	*Clerodendron colebrookianum* Walp.	Leaves, roots	Rheumatic pain, cough & asthma, hypertension
Glory lily	*Glorisa superba* L.	Root	Cancerous disease & throat problem
Jorang	*Lasia spinosa* Thaw.	Young tender shoot	Diarrhea, dysentery, piles
Satomul	*Asparagus racemosus* Wild.	Root	Urinary disorder, stomach pain
Japanese Grass	*Mikania cordata* (Burm.f.)	Leaves	Cuts & wounds, Septic sore, Jaundice, fever, bronchitis
Apang	*Achyranthes aspera* L.	Aerial parts	Cold, cough, skin disease, inflammation
Apple orchid	*Dendrobium densiflorum* Wall	Seeds	Cataract prevention, relieve throat inflammation & fatigue,
Creeping cinquefoil	*Potentilla reptans* L.	Whole plant	Ulcers, inflammations, cardiovascular & gastrointestinal problems
Shikakai	*Acacia caseia* (Linn.)	Leaves, Bark	Skin disease, hair therapy
Neem	*Azadirachta indica* A. Juss.	Leaves	Diarrhoea, dysentery, cough, asthma, tuberculosis, Ringworm, Allergy& itching
Pochotia	*Vitex negundo* L.	Leaves	Diarrhoea, dysentery, Jaundice, rheumatism & lumbago, skin disease
Noyon Tora	*Catharanthus roseus* (L.) G.Don	Flowers, fruits & leaves	Cancerous disease
Pipli	*Piper peepuloides* Roxb.	Fruits, stem and roots	Cold, cough, bronchitis, skin related disease
Butterfly pea	*Clitoria ternatea* L.	Roots	Burn, stomach disorder, urinary problem
Gakhirati bon	*Euphorbia hirta* L.	Whole plant	Skin disease-Boil, blister, inflammation
Arjun	*Terminalia arjuna*	Leaves	Rheumatism & lumbago

Table 1: contd...

	(Roxb.)		
Bahkhawr	*Eryngium foetidum* Lam. L.	Fruits, leaves	Dysentery, fever
Vako	*Thunbergia gramndiflora* Roxb.	Root	Eye disease, bone problems
Bhutmulla	*Cleome gynandra* Linn.	Tender leaves, young shoots and flowers	Ear diseases, gastro intestinal disorders
Ashiheibong	*Ficus hispida* Linn.	Fruit	Diabetes & skin disease, ringworm
Lomba	*Elsholtzia communis* Coll.	leaves and flowers	Tonsillitis, fever, cough, high blood pressure, nose bleeding, menstrual disorders
Nutgall tree/Sami tong	*Rhus semialata* Murr.	Fruits & seeds	High fever, indigestion, vomiting, allergy, food poisoning

3.2. Sericulture in North East Region

Silk is one of the finest and most beautiful natural fibres of the world and is called as "the queen of the fibres". Due to its great value and usefulness, sericulture is carried out in many parts of the world. India has second position in the production of silk. In the world, India is the only country which produces all the four varieties of silk namely, Tasar, Mulberry, Eri and Muga. There is no authentic information regarding the origin and use of silk in India, but it existed in the Ganga-Brahmaputra belts from ancient times. Out of the eight states of NE region, Assam is the largest producer of Muga and Eri silk and sixth in the production of Mulberry. Due to its golden yellow colour, softness, durability and finesse muga silk is popular in the entire country. Muga silk is obtained from the insect *Antheraea assama* which feeds on Som (*Persea bombycina*), Champa (*Michalia oblonga*), and Moyankuri (*Listea citrata*). Muga seed cocoons are available in Dibrugarh, North–Lakhimpur and Sivasagar districts of upper Assam which produce 45,000 kg of the total output and Kamrup district of lower Assam which produce 4,990 kg. In Kamrup district of Assam, there is a small village named as "Sualkuchi" where most of the people are mainly engaged in this business. Generally, two different strains of *Antheraea assama* (Muga silkworms) are observed which have two different larval colours *i.e.* green and blue (Fig. **2**).

Sericulture is mainly found in Brahmaputra valley and surrounding hill areas of Garo hills, Bhoi areas of Khasi hills and Dimapur area of Nagaland. The Eri silk is produced by the Eri silkworm, *Philosoma ricini* which feeds on Castor (*Ricinus communis*), Bas Kesseru (*Alianthus glandulosa*). Silkworms are bivoltine *i.e.* two crops in a year and moults four times in 30-35 days from hatching to mature cocoon formation. The hormone ecdysone, secreted by prothoracic gland plays a vital role in moulting. In Arunchal Pradesh, about 2,000 families are rearing this silkworm, mainly from Lohit, Tirap, East Siang, West Kameng, Lower Subansiri districts.

Blue

Green

Figure 2: Strains of Muga silkworms (*Antheraea assama*).

Bombyx mori is well known for its production of mulberry silk & its primary host plant is mulberry (*Moras alba*) leaves [25]. This host plant is cultivated widely in the plains and lower hills of NE region. Tasar silk is obtained from the insect *Antheraea* species. The silk is coppery in colour. Host plants of these silkworms are Arjun (*Terminalia arjuna*), Sal (*Sorea robusta*), Asan (*Terminalia tomentosa*).

Due to the salubrious and congenial atmosphere, Nagaland is the home of all the commercially known varieties of silkworms. Here, Eri culture is practiced as a traditional occupation. State government has established the Directorate of Sericulture to fully exploit the available potential in the state and so far, eight mulberry farms have been established in the districts of Kohima, Wokha, Zunheboto, Mokokchung districts.

Mizoram has made significant progress in sericulture development. Here, more than 4,500 families are engaged in silkworm rearing from Aizawl, Kolasib, Lungle, Variengte, Bilkhawdhlit, Thingsai, Kawnpui localities.

Like Assam, in Meghalaya all the four types of silkworms are found. It is due to the fact that all the host plants like mulberry, castor, soalu, kesseru are available in this state. There are seed farms at Shillong, Nongpoh, Tura, Ummulong and Mawdynmai.

Rearing of silkworms on large scale is assisted by State Government and Central Government agencies in the form of money, seeds, technical knowledge, insecticides *etc*. There is a special unit under State Government in each of the silk producing states. In Assam, there is Central Muga, Eri Research and training Institute (CMERTI) at Lahdoigarh, Jorhat, Assam along with their sub units and state sericulture departments to look after this industry.

4. CONSERVATION STRATEGIES FOR THE BIORESOURCES OF NORTH EASTERN REGION

In India forest management was initated right from early stages [26]. The conservation of the bio resources present in North Eastern part of India and the sustainable utilization of these resources should be given utmost priority. Emphasis must be given to programs like social forestry, community forestry, *etc*. The two conservation strategies *i.e. ex situ* and *in situ* conservation tactics should be applied for the conservation of these resources. Out of these the ex- situ conservation is the most vital component for the sustainable use and economic exploitation of these resources.

Zoos and botanical gardens are the common conventional methods of *ex situ* conservation [27]. Seed banks or germplasm banks are different ways to preserve endemic plants.

Ex situ conservation is the most vital component for the sustainable use and economic exploitation of sericulture biodiversity. Genetic variability of all the types of silkworms host plant species should be studied for documentation and utilization in breeding programs. Rearing of different types of silkworm germplasm in selected areas with conducive climate may be an alternative for *ex situ* conservation. Moreover, the wild races may be conserved *in situ* in its natural habitat.

The traditional farming systems have a key role in *in situ* conservation of plant diversity. The traditional farming systems were developed by farmers over years of experience to suit specific ecological conditions with a view to attaining stability and diversification. Since time immemorial conservation of natural resources has been an integral part of the human culture in different ways. People's participation is very important to integrate ecosystem conservation and rural development. The indigenous people are the integral part of the ecosystem. They are not only familiar with several plant/animal species in their ecosystem but also understand the ecological interrelationship of the various components of their resource.

Different branches of the central government have taken the responsibility to conserve the natural resources of North Eastern part. Botanical survey of India at Shillong and Meghalaya has taken the responsibility to conserve a huge number of plants. Apart from these the CSIR- North East Institute of Science and Technology, Jorhat, Assam have conserved quite a few species of Mushroom which are native to these regions. There is a number of different NGO's working in the north east region, actively engaged in conservation of the resources of this region.

Microorganisms represent a huge reservoir of resources providing innovative applications useful to human society. Therefore, characterization and conservation of the microbial diversity as bio resources or reservoir of diverse gene pool is of utmost priority considering the loss of habitats or ecological niches due to continuous depletion of rain forest cover and man made changes in topography. It is well recognized that the diversity of microbial communities in North- eastern region remains unexplored and uncharacterized.

5. UTILIZATION OF BIORESOURCES OF NORTH EASTERN REGION

According to a survey conducted by FAO, it is found that 40% of the world's economy depends upon the use of biological resources directly or indirectly. The raw materials such as oil, timber, herbs, minerals, *etc* are crucial for the domestic production of goods. Today our only option is to manage productivity of the resources in a sustainable manner, reducing alteration of habitats by human activities and taking into account traditional knowledge which contributes to the maintenance of ecosystem and natural resources. The North East region of India is well known for its rich flora and fauna and wealth of living ethno-medicinal traditions.

About 75-80% of the world population mainly in developing countries is still dependent upon herbal medicine for basic health care [28]. The medicinal plants used by the local tribes of the region act as a potential source for development of modern useful drugs. Nearly every plant from the smallest herb to a giant tree is a raw source for medicinal purpose to the tribesmen [29, 30]. As an example, *Clerodendrum colebrookianum*, a perennial shrub popularly known as 'Nephaphu' in Assam, 'Orematong' in Nagaland, 'Phuinum' in Mizoram 'Jarem' in Meghalaya has been reported to be useful in treatment of various ailments and diseases (anti-microbial, antihelmintic, anti-inflammatory, anti-malarial, antidiabetic, hepatoprotective, indigestion, blood pressure antihypertension, fever, ashthma), by the local tribes of the region [31].

The active components of the plant extracts can be used for the production of semi-synthetic drugs. For example some steroidal drugs like corticosteroids, sex hormones, oral contraceptives and anabolic agents are obtained from the tubers of *Dioscorea composita* and *D. floribunda*. The fruits, stem and roots of *Piper peepuloides* Roxb. are used as medicine in skin related diseases in Khasi and Jaintia Hills in Meghalaya and for cough, cold & bronchitis is Assam. Allicin an active ingredient of freshly crushed garlic shows antibacterial activity against both Gram- negative and Gram- positive bacteria [6, 15].

It has been reported that *Achyranthes aspera* is used as effective agent against treatment of urinary disorders amongst several communities in Arunachal Pradesh. Jaintia tribes use the root powder of *Asparagus racemosus* for the cure of urinary disorders and for stomach ache. *Cataranthus roseus*, is an anti-cancer drug yielding plant which shows its medicinal property against diabetes. Another very important medicinal plant, *Centella asiatica* is used against stomach disorder by different tribes and communities of North East India [32]. The crude extract of the leaves of *Cleome gynandra* is used to treat skin ulcers and ring worm infection by the Chorei tribes of Southern Assam [33]. There are so many other plant species found in this region which are medicinally important and are utilized by the local people of this region.

Another important resource of NE India is Silkworm. Sericulture is an agro based industry, and Assam is well known for production of all four varieties of silk in the country. Muga sericulture provides an excellent opportunity for the development of rural economy of the region as it is linked with the tradition and culture of local population, availability of naturally grown host plants, eco friendly production process and expertise in rearing, reeling and weaving process. Due to its suitable environmental conditions India is well known for local species of silkworms that contain many resistant genes. Many disease resistant hybrids of silkworm can be produced with these genetic resources. Another approach is the use of molecular markers as tool for inheritance studies of complex traits. Other molecular techniques like PCR based molecular marker, random amplified polymorphic DNA (RAPD) had been used for the identification and selection of good quality silkworm for production of healthy and disease resistant hybrids. Since time immemorial medicinal plants have been used as remedies for human diseases but they have the potential to control many insect diseases. An attempt was made successfully against the flacherie disease of silkworm by using extract of some medicinal plants available in NE India [34, 35].

In this new era of science people have to explore natural bio-resources and study the complexity and vitality of bio-resources as well as their innumerable medicinal, agricultural and industrial applications. The continual and dynamical development of faster molecular techniques, together with the advancement of methods to deal with the exponentially increasing amount of data based on the natural resources of this region are expanding our capacity for the analysis of plant, insect and microbial communities from an unlimited variety of habitats and environments. In this new decade successful assessment of biodiversity is possible with the consensus approach of utilizing the molecular genetic techniques along with the conventional methods.

ACKNOWLEDGEMENTS

Dr. B. G. Unni, Chief Scientist & Area Coordinator (Biological Sciences/Biotechnology) is thankful to Department of Science & Technology, Govt. of India, Department of Biotechnology, Central Silk Board, Ministry of Environment & Forests, DRDO and Council of Scientific & Industrial Research for research grants. The research contribution of Dr. (Mrs) Arundhati Choudhury, Dr. Arijit Guha, Dr. DK Khanikor, Dr. Puthul Saikia, Dr. Karthik Neog, Dr. Mohan Hazarika, Dr. (Mrs) Jyotsana Kumari Sharma, Dr. (Ms) Archana Bora, Dr. Manoj Singha and Ms. Yellena Kakoty are greatly acknowledged.

CONFLICT OF INTEREST

The author(s) confirm that this chapter contents have no conflict of interest.

REFERENCES

[1] Dirzo R, Raven PH. Global state of biodiversity and loss, Annual Review Environment and Resources 2003; 28: 137-167.

[2] Hosetti BB. Endangered Plants, Concepts in Wildlife Management, 2nd revised and enlarged edition 2005; 11-18.

[3] Tandon P and Kumaria S. Biodiversity of North East India and its Conservation, Biodiversity and its significance, edited by Tandon P, Abrol AP, Kumaria S. 2007; pp. 157-160.

[4] Nath SC. Plant Folk-Medicines and Medicinal Plants of North-East India, published by CSIR-NEIST 2011; pp. 3-166.

[5] Myers N, Mittermeier RA, Mittermeier CG, Fonseca GAB, Kent J. Biodiversity hotspots for conservation priorities, Nature 2000; 403: 853-858.

[6] Unni BG, Wann SB, Borah A, Devi B. Potential of medicinal, aromatic and essential oil yielding plant resources and prospects of their commercial cultivation in North-East India, Ethnoforestry: The Future of Indian Forestry, edited by S.C Tiwari 2009(a); pp.79-98.

[7] Champion HG, Seth SK. A Revised Survey of Forest Types of India, Govt. of India Press 1968; pp. 404.

[8] Chakraborty R, De B, Devanna N, Sen S. North-East India an Ethnic Storehouse of Unexplored Medicinal Plants, Journal of Natural Products and Plant Resources 2012; 2(1): 143-152.

[9] Chatterjee S, Saikia A, Dutta P, Ghosh D, Panggiang G and Goswami AK. Biodiversity Significance of North East India: for the study on Natural Resources, Water and Environment Nexus for Development and Growth in North Eastern India 2006; 13.

[10] Mao AA, Hynniewta TM, Sanjappa M. Plant wealth of Northeast India with reference to ethno botany, Indian Journal of Traditional Knowledge 2009; 8: 96-103.

[11] Chakravorty R, Dutta P, Ghose J. Sericulture and traditional craft of silk weaving in Assam, Indian Journal of Traditional Knowledge 2010; 9 (2): 378-385.

[12] Singh EJ, Singh NKS, Singh NR. Biodiversity conservation and natural resources in North East India-with special reference to Manipur, NeBIO 2009; 1 (1); 42-47.

[13] Unni BG Murthy USN. Sericulture and viable technologies for better silk production, Current Science 2008; 95(1): 14.

[14] Verma PS, Agarwal VK. Wildlife Management in India, Environmental Biology 2005; 1: pp. 538-550.

[15] Unni BG, Goswami M, Kakoty Y, Bhattacarjee M, Wann SB, Rajkhowa G, Das S, Devi BR, Das AC. Indigenous knowledge of silkworm cultivation and its utilization in North Eastern region of India, Indian Journal of Traditional Knowledge 2009(b); 8(1): 70-74.

[16] Lisam KS. Geography, Fauna and flora, Encyclopaedia of Manipur 2011; 3: pp 55-67.

[17] Arrawatia ML, Tambe S. Biodiversity of Sikkim: Exploring and conserving a global hotspot, Information and public relations department, Govt. of Sikkim 2011.

[18] Bhupathy S, Kumar SR, Thirumalainathan P, Paramamamdham J, Lemba C. Wildlife exploitation: a market survey in Nagaland, North-eastern India, Journal-Tropical Conservation Science 2013; 6 (2):241-253.

[19] Trivedi PC. Some sacred trees and their medicinal uses, Medicinal plants: Traditional Knowledge 2006; 1: pp-236-243.

[20] Khongsai M, Saikia SP, Kayang H. Ethnomedicinal Plants used by different tribes of Arunchal Pradesh, Indian Journal of Traditional Knowledge 2011; 10(3): 541-546.

[21] Dominic R, Ramanujam. Traditional Knowledge and ethno botanical uses of piscicidal plants of Nagaland, North east India, Indian Journal of Natural Products and Resources 2012; 3(4): 582-588.

[22] Sangtam TL, Jamir NS, Deb CR, Jamir S. A Study on the Medicinal Plants used by the Sangtam Naga Tribe in Kiphire District, Nagaland, India, International Journal of Ayurvedic and Herbal Medicine 2012; 2(2): 267-275.

[23] Buragohain J. Ethnomedicinal Plants Used by ethnic communities of Tinsukia District of Assam, India, Recent Research in Science and Technology 2011; 3(9): 31-42.
[24] Das T, Mishra SB, Saha D, Agarwal S. Ethnobotanical Survey of Medicinal plants used by Ethnic and Rural people in Eastern Sikkim Himalayan region, African Journal of Basic and Applied Sciences 2012; 4 (1): 16-20.
[25] Murthy VNY, Ramesh HL, Munirajappa. Impact of feeding selected mulberry germplasm varieties on silkworm *Bombyx mori* L. through bioassay techniques for commercial exploitation, Asian Journal of Natural & Applied Sciences 2013; 2 (4): 56-64.
[26] Saldanha IM. Colonialism and professionalism: A German forester in India, Environment and history (2) 1996; 2: 195-219.
[27] Fetene A, Yeshitela K, Desta H. Approaches to Conservation and Sustainable Use of Biodiversity-A Review, Nature and Science 2012; 10(12): 51-62.
[28] Kamboj VP. Herbal medicine, Current Science 2000; 78: 35-51.
[29] Unni BG, Singh R, Neog K, Bhattacharjee M. Isolating muga silkworm (*Antheraea assamensis* Helfer) genomic DNA without liquid nitrogen suitable for marker studies, African Journal of Biotechnology 2011(a); 10(55): 11365-11372.
[30] Unni BG, Dowarah P, Wann SB, Rao PG. Muga Heal-*Terminalia chebula* based Bioformulation as an anti-flacherie agent and a silk fiber enhancer, Journal Science & Culture 2011(b); 77(11-12): 456-460.
[31] Kalita J, Singh SS, Khan ML. *Clerodendrum colebrookianum* Walp., A potential folk medicinal plant of North East India, Asian Journal of Pharmaceutical and Biological Research 2012; 2: 256-261.
[32] Sajem AL, Gosai K. Traditional use of medicinal plants by the Jaintia tribes in North Cachar Hills district of Assam, Northeast India, Journal of Ethnobiology and Ethnomedicine 2006; 2: 1-7.
[33] Choudhury S, Sharma P, Choudhury MD, Sharma GD. Ethnomedicinal plants used by Chorei tribes of Southern Assam, North Eastern India, Asian Pacific Journal of Tropical Disease 2012; 2: 141-147.
[34] Neog K, Singh HR, Unni BG, Sahu AK. Analysis of genetic diversity of muga silkworm (*Antheraea assamensis* Helfer; Lepidoptera: Saturniidae) using RAPD-based molecular markers, African Journal of Biotechnology 2010; 9(12): 1746-1752.
[35] Singh R, Unni BG, Neog K, Bhattacharjee M. Sodium Dodecyl Polyacrylamide gel electrophoresis (SDS-PAGE) and random amplified polymorphic DNA (RAPD) based genetic variation studies in eri silkworm (*Samia cynthia ricini*) Lepidoptera: Saturniidae), African Journal of Biotechnology 2011; 10(70): 15684-15690.

CHAPTER 7

Sacred Groves of Himachal Pradesh: A Traditional Community Practice of Biodiversity Conservation and Testimony of Religious Beliefs

Vandana Thaplyal* and Ashok Verma

WWF-India, Field Office, Bishop Cotton School, Shimla, Himachal Pradesh, India

Abstract: The harmony between human and nature is maintained through civilizations by practicing reverence, recognition and appreciation. Nature has been worshipped and the anthropogenic arrogance of excessive use of natural resources has been prohibited or else mankind has faced the vagaries of nature. The communities have considered biodiversity as one of the important aspects of sustainability of mankind and have marked landscapes with mythological importance as sacred groves. The patches of dense vegetation or natural areas which have ethnic and religious values are known as sacred groves. These sacred groves are protected by communities through local taboos and sanctions. In sacred groves a range of traditions and cultural values are manifested. Traditionally, wild species were protected and conserved in the sacred groves with often strict penalties for breaking customary laws on the use of resources within the groves. In Himachal Pradesh, the sacred groves are called *Dev Van* or *Devta Ka Jungle* and are dedicated to a particular deity. The paradigm shift in customary practices and modernization has threatened the future of these sacred groves. There is a need to exercise ethical concern towards these landscapes of rich biological diversity. A holistic approach, cutting across multidisciplinary subjects has been adapted to document sacred groves. A total of 514 sacred groves have been documented considering biodiversity, culture, religion and history of the landscapes.

Keywords: Biodiversity, communities, endemic plants, GPS, Himachal Pradesh, medicinal plants, sacred grove.

1. INTRODUCTION

The preservation of landscapes has been an integral part of human civilization all over the world. Management of natural resources has been traditionally adopted because of the dependency of mankind over it. Man has always been worshipping nature which has proved to be a very effective tool for conservation. The natural sites such as rivers, lakes, forests, gardens, groves *etc.* have been

***Corresponding author Vandana Thaplyal:** WWF-India, Field Office, Bishop Cotton School, Shimla, Himachal Pradesh, India; Tel: 09816026295; E-mail: vandanathaplyal@gmail.com

Laladhas K.P., Oommen V. Oommen and Sudhakaran P.R. (Eds.)

institutionalized by attaching values with an intention to make collective management easy and sustainable [1]. The various religious philosophies have contributed significantly in the conservation of forests, biodiversity and landscapes by promulgating customary practices and beliefs. Some live examples of traditional conservation practices still exist in the form of sacred groves [2]. Sacred groves are one of the most valuable legacies of nature conservation which have been completely immune from human interference on grounds of religious beliefs. The folklore associated with the sacred groves play an important role in confirming such beliefs and points blessings for good behaviour and heavy punishment for infidel [3]. These patches of vegetation or natural areas which have ethnic and religious values are known as sacred groves [4]. Sacred groves are tract of virgin forests, harbouring rich diversity and protected traditionally by the local communities. These patches of forests are left untouched because of social fencing [5]. The sacred groves are usually dedicated to local deities and are associated with beliefs, taboos and sanction systems. In sacred groves a range of traditions and cultural values are manifested. These age old institutions are surviving all over the world and conserving natural resources [6]. The landscapes with sanctity and mythological importance have been marked as sacred groves. Traditionally, wild species were protected and conserved in the sacred groves with often strict penalties for breaking customary laws on the use of resources within the groves. The importance of sacred groves in socio-religious life as well as livelihood security has been felt by the indigenous communities from time immemorial. In India the origin of sacred grove is considered to be pre-vedic. Setting aside patches of forest land as sacred groves has been the practice by communities for centuries in India. There are sacred groves in many parts of India which are known by several names such as Kavu in Kerala, Devaravana in Karnataka, Devgudi in Maharashtra and Kenkaris in Rajasthan [7]. In Himachal Pradesh, the sacred groves are called *Dev Van* or *Devta Ka Jungle* and are dedicated to a particular deity (Fig. **1**).

Sacred groves are the repository of endemic species and can be regarded as the remnant of the primary forest left untouched [8]. Sacred groves are not only the sacred ecosystems but also a product of socio-ecological philosophy [9]. These are the home of local flora, fauna, gene pool and mini biosphere reserve [10]. The sacred groves represent climax vegetation of rare and endangered species of plants [11]. They possess a rich diversity of medicinal plants and provide suitable habitat for their sustainable natural regeneration [12]. The sacred groves are well developed ecosystems and represent biological and cultural heritage of the area. They protect plants as well as animals, birds, insects and microorganisms representing regional biodiversity [6]. The composition of sacred groves represents magnificent specimens of biodiversity and these forest patches are devoid of human interference [13-15]. The paradigm shift in customary practices, changing demographics and modernization has threatened the future of these sacred groves.

Figure 1: Jamdagini Rishi Temple & sacred grove in Gadsa valley, Kullu.

2. SACRED GROVES OF HIMACHAL

The present study is based on the field survey by using *global positioning system* (GPS), topo sheets and village map to know the location of sacred groves. All the 12 districts of Himachal Pradesh were surveyed. Floral survey was done by collecting samples and photographs. A holistic approach, cutting across multidisciplinary subjects has been adopted during the questionnaire survey of the sacred groves. The communities have been consulted to record historical, religious and cultural background as well as sanctions and taboos associated with the sacred groves. The ownership of the sacred grove is also recorded. Threats to the biodiversity of sacred groves were also taken into account to find out the solutions to conserve biodiversity. The main focus of the study was to document woody flora. The plants of religious importance have also been recorded. Dominant plant species and other secondary species of sacred groves have been recorded. The faunal information has been gathered through direct sighting and from local people. The surveys of the villages adjacent to sacred grove have also been done to know the concerns about degradation and conservation of sacred groves.

Himachal Pradesh, a North West Himalayan State is located between 30° 22'40" and 33°12'40"N latitude and between 75°47'55" and 79°4'20"E longitude. The state is bordered by Jammu and Kashmir in the North, Punjab and Haryana in the South West and the state of Uttarakhand in the South East. The altitudinal range of the state is between 450 m in the foothills to 6500 m in the high hills. The state has mixture of agro climatic zones from sub tropical like Solan, Hamirpur, Una, Bilaspur parts of Chamba, Kangra, Shimla and Sirmaur districts to temperate region like parts of Kullu, Chamba, Shimla, Mandi, Sirmour and Kangra districts and cold deserts of Lahaul - Spiti and Kinnaur in the trans Himalayas (Fig. **2**).

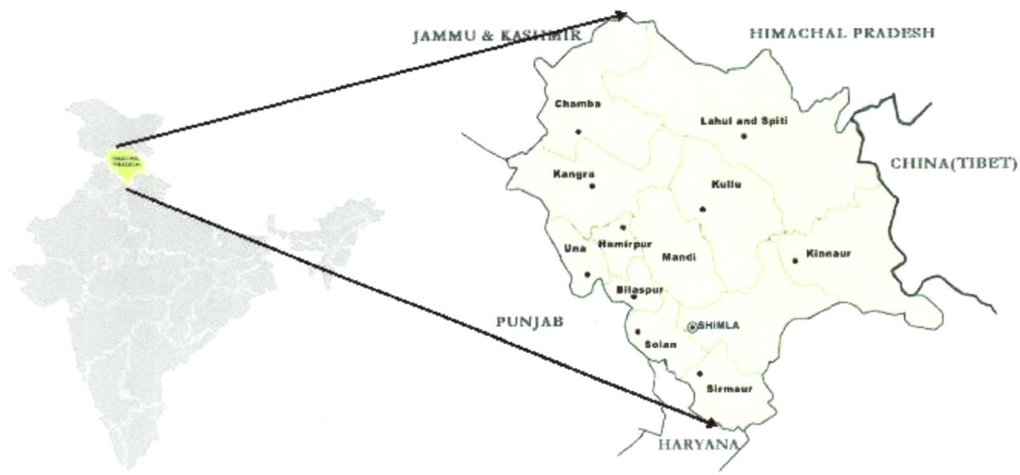

Figure 2: Study area: Himachal Pradesh.

A total of 514 sacred groves have been documented in 12 districts of Himachal Pradesh, considering biodiversity, culture, religion and history of the landscapes. Shimla district represents highest number of sacred groves *i.e.* 171, followed by 108 sacred groves in Kullu, 40 in Solan, 39 in Mandi, 28 in Chamba, 26 in Sirmour, 24 in Kangra, 19 in Hamirpur, 17 in Kinnaur, 16 in Bilaspur, 15 in Una and 11 in Lahaul & Spiti (Figs. **2** and **3**).

The altitudinal range of sacred groves in study area is between 385 m asl to 3674 m representing diversified flora. The sacred grove of Baba Rudranand Temple of Naari village of Una district is in lowest elevation and the sacred grove around Nako lake of Kinnaur district is in highest elevation. Study reveals there are 27 species of trees which are present as dominant species in subtropical zone, 109 sacred groves have *Pinus roxburghii* as dominant species followed by 15 sacred groves with *Acacia catechu* as dominant species. The temperate zone has *Cedrus deodara* as dominant species in 285 sacred grove followed by *Quercus leucotrichophora* as dominant species in 30 sacred groves (Fig. **4**). Other tree species shrubs and ground flora are also present. The plants of ethnobotanical importance are also conserved in the sacred grove.

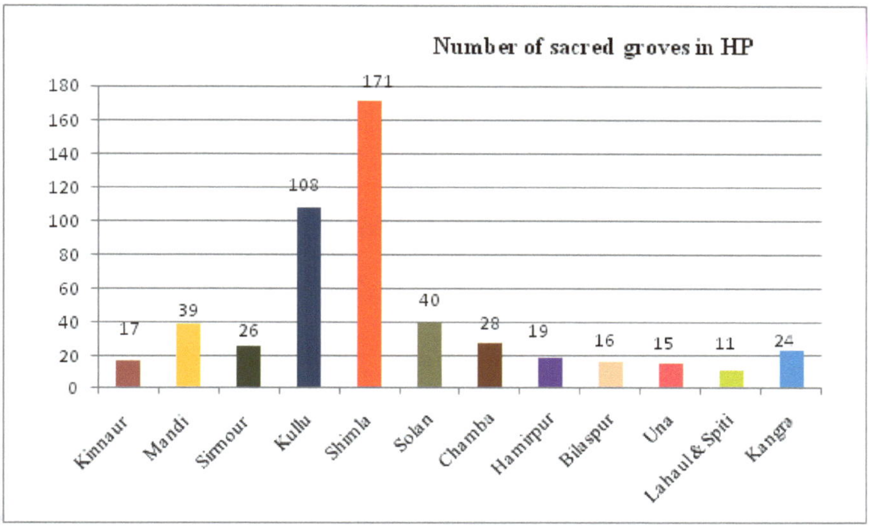

Figure 3: Number of sacred groves in each district of Himachal Pradesh.

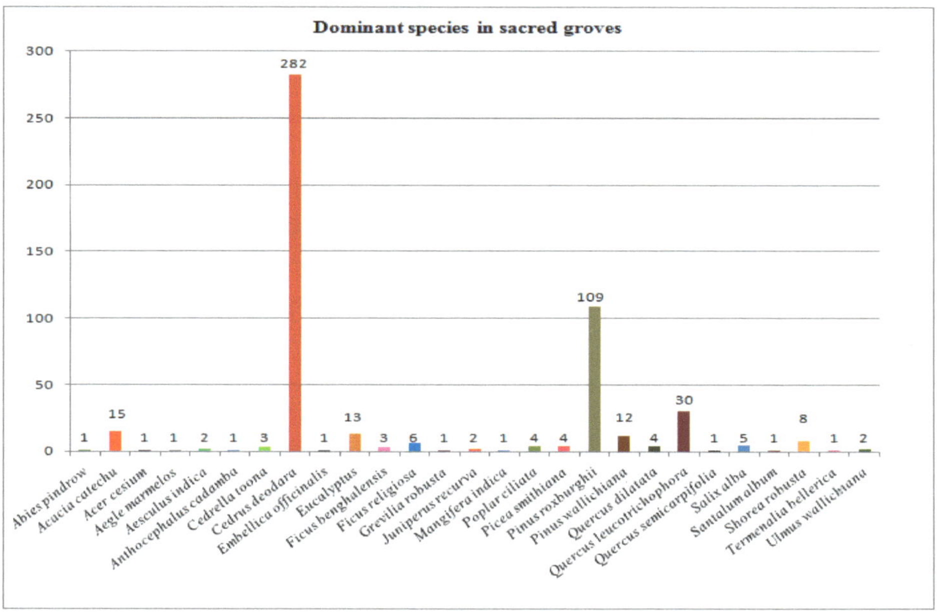

Figure 4: Dominant species in sacred groves.

The size of the sacred groves documented varies from 0.02 ha to 72.8 ha. The sacred grove associated with Mahavir Termple of village Johana in Sirmour district has the lowest area while the sacred grove associated with Kamrunag Temple of Naraili village of Mandi district has the largest area (Fig. **5**).

Figure 5: Kamrunag Temple of Naraili village, Mandi.

The management of sacred grove is either with temple committee or with Forest Department. There are sacred groves owned by the local deity and managed by the temple committee. 188 sacred groves are managed by Forest Department, 309 sacred groves are in the name of the local deities and are managed by the temple committee. 17 sacred groves are partly in the name of the Forest Department and partly in the name of the deity but are managed jointly. The sacred grove has either a temple or a seat of the deity with a well organized temple committee. The management of the temple committee has *Pujari* (priest), *Gur/ Mali,* (messenger of deity), *Bhandari* (store keeper), *Kardar* (manager) and *Sadasya* (members) who are responsible to do various work assigned to them. The local fairs and festivals are organized within the temple premises and the revenue collected is used for the maintenance of the temple. The wood from the sacred grove is used only for construction of the temple or as a fuel during ritual in the temple premises. No one is allowed to take any sort of forest produce from the sacred grove. In Kullu district, there are few sacred groves where people can enter bare foot and are not allowed to extract even dead wood. Leather goods and metal tools are also prohibited in few sacred groves. There is much folklore associated with the deity of the sacred grove.

It has been found that all the species within the sacred groves are not considered sacred. The sacred grove may have or may not have sacred species of plants. *Cedrus deodara, Pinus roxburghii, Ficus religiosa, Ficus benghalensis, Prunus cerasoides etc.,* are considered to be sacred. The rich floral diversity is a good habitat for faunal diversity which includes leopard, wild boar, barking deer in the subtropical region; black and brown bear, porcupine, fox, ghoral, Himalayan thar, musk deer in temperate region and snow leopard, ibex and bharal in the trans Himalayas. Himalayan birdlife and pheasants in particular are well known in the state. The dense vegetation of sacred groves trap rain water and replenish ponds, springs, wells, streams and *bowaris*.

CONCLUSION

The practice of conservation of plant species in the form of sacred grove represents the ancient ethos of the community of Himachal Pradesh. It has been estimated that most of the sacred groves in Himachal Pradesh are approximately 500 years old. There is a need to exercise ethical concern towards these landscapes of rich biological diversity so that the sanctity of the sacred grove is maintained. Many sacred groves are of small size but have religious and aesthetic importance. The sacred groves with vast area have vital influence on biodiversity and local climate. The sacred groves are maintained through *in situ* conservation and natural regeneration. No afforestation, enrichment or extension of sacred grove has been noticed during the study. No legal status is given to the sacred grove; however they occur in reserve forests, demarcated forests, unclassified forests and common lands. The ever increasing demand of modern society has seen ruthless destruction of natural resources to meet their greed. This has resulted in minimizing the density and encroachments in many of the sacred groves. Forest fire and land slides are other threats to these pristine landscapes. The ecology in the study area is adversely affected due to anthropogenic and development activities. The study of sacred groves can help in planning and developing strategies for future conservation of the ecosystems and species. It has been felt that the sacred groves can be conserved and managed better through institutional building and providing legal status to the sacred groves.

ACKNOWLEDGEMENTS

The authors are thankful to WWF- India for providing support to implement the project. The financial support given by G.B. Pant Institute of Himalayan Environment and Development, Almora is acknowledged. We would also like to thank local communities of Himachal Pradesh and Forest Department for providing information regarding sacred groves. Technical support from Dr. Sejal Worah, Dr. Archana Godbole, Dr. S.S. Samant, and Dr. G.S. Goraya is acknowledged.

CONFLICT OF INTEREST

The authors confirm that this chapter contents have no conflict of interest.

REFERENCES

[1] Pruthi I, Burch Jr W. A Socio-Ecological Study of Sacred Groves and memorial parks: Cases from USA and India. International Journal of Environmental Science and Engineering 2009; 1 (1): 7-14.
[2] Kala C P. Traditional ecological knowledge, sacred groves and conservation of Biodiversity in the Pachmarhi biosphere Reserve of India. Journal of Environment Protection 2011; (2): 967-973. http://www.SciRP.org/journal/jep.
[3] Vartak VD, Gadgil M. Studies on Sacred Groves along Western Ghats in Maharashtra and Goa: Role of Beliefs and Folklores. Contribution to Ethnobotany to India. Scientific Publishers, New Delhi 1990; pp. 329-335.
[4] Mohanta R K, Bhardwaj A K , Adhikari B S , Mathur P K . Distribution a n d Conservation Status of Scared Groves (SGs) in Garo Hills, Meghalaya. Indian Forester 2009; 135 (12):1627-1649.
[5] Khan ML. Center for the Conservation of Biodiversity: Sacred Groves in India. XII World Forestry Congress, Quebec City, Canada 2003. http://www.fao.org/docrep/ARTICLE/WFC/XII/0509-A1.HTM.
[6] Godbole A, Sarnaik J. Tradition of sacred groves and communities contribution in their contribution. Applied Environment Research Foundation, Pune (2004).
[7] Khan ML, Khumbongmayum AD, Tripathi RS. The sacred Grove and their Significance in Conserving Biodiversity: An Overview. International Journal of Ecology and Environmental Sciences 2008; 34 (3): 277-291.
[8] Anthwal A, Sharma RC, Sharma A. Sacred Groves: Traditional way of Conserving Plant Diversity in Garhwal Himalaya, Uttaranchal. The Journal of American Science 2006; 2 (2): 35-43.
[9] Anubhav Singh K, Rastogi A, Singh V. Sacred Groves in India: Celebrating Sanctity of Life through Biodiversity Conservation. waswc.soil.gd.cn/./Vir%20Singh/Sacred%20groves%20 in%20India,%20 Anubhav%20et%20al%20080312.doc.
[10] Deepamol PS, Khaleel KM. Comparison of Floral Diversity in Fresh water and salt water wetlands Sacred Groves of Kannur District. Journal of Phytology 2009; 1(5): 299-301.
[11] Singh H, Husain T, Agnihotri P. Haat Kali sacred grove, Central Himalaya, Uttarakhand. Current Science 2010; 98 (3): 290.
[12] Khumbongmayum AD, Khan ML, Tripathi RS. Ethnomedicinal plants in the sacred groves of Manipur. Indian Journal of traditional knowledge 2005; 4(1): 21-32.
[13] Bhagwat SA, Cheppudira G, Kushalappa Paul HW, Brown NC. A landscape Approach to Biodiversity Conservation of Scared groves in the Western Ghats of India. Conservation Biology 2005; 19 (6): 1853-1862.
[14] Yadav S, Yadav JP, Arya V, Panbhal M. Sacred Groves in Conservation of Plant Biodiversity in Mahendergarh district of Haryana. Indian Journal of Traditional knowledge 2010; 693-700.
[15] Ambaashan M, Parthasarthy N, Padmavathy A. Ethno-floristic Survey in sacred groves, Pudukottai district, Tamil Nadu-India. Journal of Medicinal Plants Research 2011; 5 (3): 439-443.

CHAPTER 8

Population Density and Conservation Status of Indian Gaur (*Bos gaurus gaurus*) in Trishna Wildlife Sanctuary, Tripura, Northeast India

Mary P.P.[1,*], Debbarma J.[1], Kumar Awadhesh[1], Gupta A.K.[2]

[1]North Eastern Regional Institute of Science and Technology (NERIST) (Deemed University), Arunachal Pradesh, India and [2]Dept. of Environment & Forest, Govt. of Tripura, P. O. Kunjaban, Agartala, Tripura (W), India

Abstract: Indian gaur or Indian bison (*Bos gaurus gaurus*) is a wild bovid. It is categorized as vulnerable in the IUCN Red list of threatened species, 2009 and listed in schedule-1 of Indian wildlife (protection) act, 1972. The present population density and conservation status of gaur in Tripura was studied. Trishna wildlife sanctuary, popularly known as "Eden of Bison", is situated in the South Tripura district. It is home to five primate species, namely, Spectacled langur (*Trachypithecus phayrie*), Rhesus macaque (*Macaca mulatta*), Pig-tailed macaque (*Macaca nemtsetrina*) Capped langur (*Trachypithecus pileatus*) and Hoolock gibbon (*Hylobates hoolock*). Physical characters and feeding behaviour of gaur are recorded. Direct counting of stratified block count method was followed to assess the population strength of gaur. The whole sanctuary was stratified into three habitats, *in situ*., dense forest, degraded forest and bushy forest. Counts of sample blocks of the three strata were recorded and total population of the sanctuary was found to be 255±139. Stratum-wise population density was calculated to be 0.99, 1.67 and 1.88 for dense forest, degraded forest and bushy forest respectively. Many conservation measures are taken up by the Forest department to protect this species and its habitat. However, some anthropogenic threats and adverse effects of global warming on gaur are recorded. Further measures are recommended to conserve Indian gaur and its habitat for the benefit of future generations.

Keywords: Bison, conservation strategies, hunting, Indian gaur, indigenous knowledge, poaching, population density, vulnerable, wild life sanctuary.

1. INTRODUCTION

Indian gaur or Indian bison (*Bos gaurus gaurus*), (Fig. **1**) is the heaviest and the most powerful of all wild cattle. They are categorized as Vulnerable in the IUCN

***Corresponding author Mary P.P.:** North Eastern Regional Institute of Science and Technology (NERIST) (Deemed University), Arunachal Pradesh, India; Tel: +91-0360-2257401 – 2257415; Extn. 6228 (work); Mobile: +91 9612422691; Fax: 91-0360-225853; E-mail: purakat@yahoo.co.in

Laladhas K.P., Oommen V. Oommen and Sudhakaran P.R. (Eds.)

Red List of Threatened species. They are listed in Schedule-I of the Indian Wildlife, (Protection) Act of 1972. In the past gaur population was estimated to be 12000 – 22000 in evergreen forests of Western Ghats in particular National Parks of Karnataka, TamilNadu and Kerala [1]. In northeast Indian states of Mizoram and Tripura, Indian gaur is reported to cross international border to Bangladesh [2]. Occurrence of gaur in Arunachal Pradesh was reported [3]. However, the current status of gaur population in North East India is poorly known. Hence a study was conducted during January-April'2011 to assess the present population density and conservation status of gaur in Trishna wildlife sanctuary, Tripura.

Figure 1: Bos gaurus gaurus.

2. STUDY AREA AND METHOD

2.1. Eden of Bison

Trishna wild life sanctuary, "Eden of Bison"- covers 197 sq. km. in area, located in South Tripura District (Latitude: - 23° - 12' North to 23° - 32' and Longitude: - 91°- 15' East to 91° - 30') (Fig. **2**). The sanctuary has rich bio-diversity and three eco-tourism zones. Prominent flora in the habitat of gaur comprises "kaillai" bamboo (*Oxytenanthera nigrociliata*), congo grass, tail grass *etc*. The sanctuary has a unique faunal diversity comprising of several migratory birds and indigenous mammals especially primates. It is home to five primate species, namely, Spectacled langur (*Trachypithecus phayri*), Rhesus macaque (*Macaca mulatta*), Pig-tailed macaque (*Macaca nemestrina*), Capped langur (*Trachypithecus pileatus*) and Western Hoolock gibbon (*Hoolock hoolock*).

Map 1. Trishna wild life sanctuary

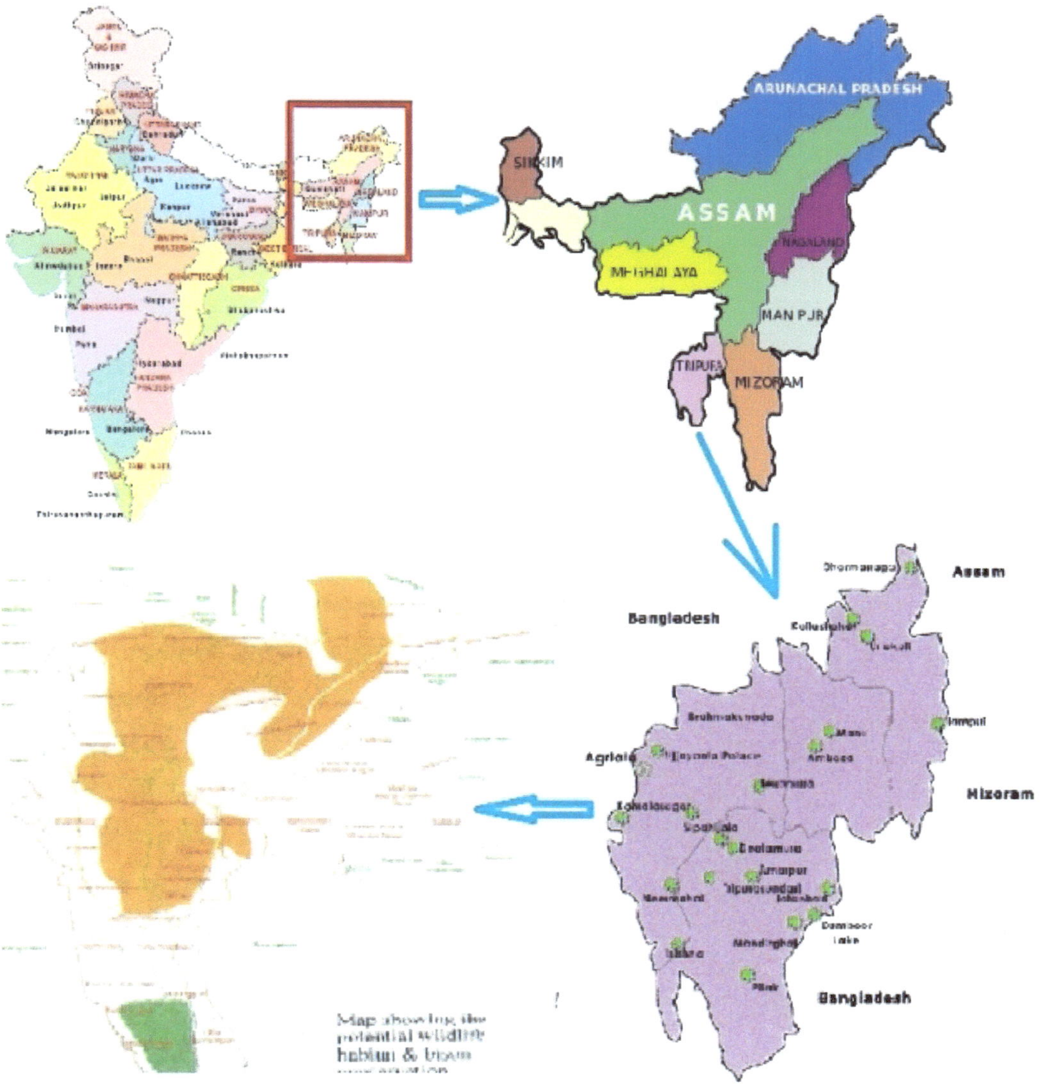

Direct stratified block count method was adopted to estimate the total population of Indian bison in the sanctuary. The study was conducted in three strata, which are dense forest, degraded forest and bushy forest. Based on the direct sighting of Indian bison in sample blocks, the total population present in the entire area of the sanctuary was estimated [4]. Out of total area of sanctuary (197.7km^2) dense forest stratum was 122.57 km^2 (62%) degraded forest stratum was 35.59 km^2 (18%) and bushy forest stratum was 39.54 km^2 (20%).

Table 1: Population density of Indian gaur in sample blocks of three strata

Stratum	**Dense forest** Sample Sample Block- 1 block- 2 ($6km^2$) ($6km^2$)	**Degraded forest** Sample Sample block- 3 block- 4 ($3km^2$) ($3km^2$)	**Bushy forest** Sample Sample block- 5 block- 6 ($3km^2$) ($3km^2$)
No. of Gaur sighted	7 2	4 3	7 4
Density-sample Block-wise	1.166 0.333	1.333 1.0	2.333 1.333
Density Stratum-wise	0.75	1.17	1.83
Standard deviation	0.589	0.235	0.707
Standard error	0.416	0.166	0.50
Confidence Limit 95%	0.92	0.37	1.1

2.2. Population Density

Stratum-wise population density was 0.75, 1.17 and 1.83 in dense forest, degraded forest and bushy forest respectively (Table **1**). Hence the stratum-wise population of gaur in the sanctuary is estimated as follows:-

Dense forest ----------- (0.75±0.92) x122.57 = 92±113

Degraded forest ------- (1.17±0.37) x35.59 = 42± 13

Bushy forest ----------- (1.83±1.1) x39.54 = 72± 43

Total =206±169

It means the population of *Bos gaurus* in that area as 95% situation will be either 169 or 206. Census conducted by forest department recorded total gaur population of around 92, 100, 110 and 130 gaurs in years 2007, 2008, 2009 and 2010 respectively. When compared to the previous years' census data there is an increasing trend of gaur population growth in Trishna wildlife sanctuary. Stratum-wise density was estimated and overall density was 1.25 animals per sq. km. The highest density (2.33 per sq. km.) was recorded in bushy forest area followed by 1.66 per sq. km. in Dense forest and 1.33 per sq. km. in Degraded forest area.

3. CONSERVATION STRATEGIES FOR GAUR

Increasing trend of gaur population, is the result of various management strategies taken by forest department such as plantation of food-plants (*Oxytenanthera nigrocilliate, Brachiaria ruziziensis),* creation of rain-water harvesting structure, protection activities, infrastructural development for sanctuary, observation and protection and installation of six lightning conductors to save the animals from the menace of being struck by lightning. Participation of local people is encouraged by arranging awareness activities among people around the sanctuary areas to save wildlife. Eco-tourism development activities enhances socio-economical upliftment of people living around the sanctuary by providing them work, irrigation facilities, alternate source of income as tour guides, running tourist lodges *etc.* Forest department provides salt which provides physical immunity to gaur.

4. CONSERVATION STRATEGIES TO BE ADOPTED

The following measures are recommended to conserve Indian gaur and bio-diversity of its habitat for the benefit of future generations:

4.1. Monitoring of Hunting and Poaching

Strengthening protection measures to check both subsistence and commercial hunting of Indian bison should be the first priority to ensure that exploitation of natural resources by the local communities for their sustenance needs do not exceed the carrying limits.

4.2. Awareness and Education

Second most important aspect to this effect is to conduct special awareness programmes such as video film, power point slide presentation, posters, banners, pamphlets *etc.*, developed in the local language, in a way easily understandable by the local people. They should be educated about the ecological role played by Indian bison in maintaining forest ecosystem.

4.3. Increasing the Strength of Staff

The strength of staff must be increased so that the whole forest area can be supervised in minimum period of time.

4.4. Utilization of Indigenous Knowledge

Local people with indigenous knowledge about the species and area should be directly or indirectly involved in census and conservation activities to get more cooperation in implementation of conservation action.

ACKNOWLEDGEMENTS

We are very much thankful to PCCF (Wildlife and Biodiversity), Department of Forest, Govt. of Tripura for providing necessary permission to carry out field work. We are also thankful to head of Department, Forestry, NERIST and field staffs of sanctuary for administrative support and their kind help during the data collection respectively.

CONFLICT OF INTEREST

The authors confirm that this chapter contents have no conflict of interest.

REFERENCES

[1] Ranjitsinh MK. Beyond the Tiger. Portraits of Asian wildlife. New Delhi, Birajbasi Printers India 1997.

[2] Choudhury A. Distribution and conservation of the Gaur, *Bos gaurus* in the Indian Subcontinent. Mammal Review 2002; 32: 199-226.

[3] Choudhury A. The gaur, *Bos gaurus* in Dibang Valley district of Arunachal Pradesh. Journal of the Bombay Natural History Society 1999; 96(2): 311-313.

[4] Singh SK. Text book of wild life management, IDBC publisher 2006; pp.172-178.

CHAPTER 9

Therapeutic Value of Milk of Vechur Cow, the Indigenous Breed of Kerala Listed as Critical by FAO

Mohankumar Chinnamma*, Anisha Shashidharan and Salini Bhasker

SCMS Institute of Bioscience and Biotechnology Research and Development, Management House, South Kalamassery, Cochin, Kerala, India

Abstract: *Bos indicus*, (Vechur breed), being an indigenous cow of Kerala needs greater attention for its survival since the breed is facing extinction. Due to the lack of appropriate scientific studies for highlighting the therapeutic potential of the milk of Vechur cow and its unique medicinal properties, this critical breed has not gained sufficient and required acceptance among farmers and it became a negative factor for the conservation of this domestic animal. Most of the traditional concepts associated with Vechur cow like disease resistance, medicinal property of milk have remained as mere field observations since the authenticity of these features has not yet been proven successfully with solid evidence. Hence the existence of Vechur cow in the livestock industry is still a questionable matter. The data of the present study on the structural and functional features of the lactoferrin gene of the Vechur cow demonstrates SNPs, amino acid substitutions, binding pockets and the secondary and tertiary structure of the gene compared to *Bos taurus*. Along with the genomic features, the gene was successfully expressed for the production of recombinant Bovine lactoferrin protein (rBLf) in eukaryotic system using a Pore2 vector and CAMV^{35}S promoter. The bactericidal property exhibited by rBLf protein in suppressing the growth of both gram positive and gram negative bacterial strains *viz. E.coli, Streptococcus* sps, *and Staphylococcus aureus* has highlighted the relevance of lactoferrin protein of Vechur cow milk in drug industry.

Keywords: Amino acid, anti-inflammatory, antimicrobial, antitumor, bactericidal property, *Bos indicus, Bos Taurus,* Bovine Lactoferrin, disease resistance, *E. coli*, glycoprotein, medicinal properties, milk, nucleotide polymorphism, nutrition, potential immunostimulant, *Staphylococcus*, *Streptococcus,* therapeutic value, Vechur cow.

1. INTRODUCTION

India has one of the largest cattle populations in the world coming under the

***Corresponding author Mohankumar Chinnamma:** SCMS Institute of Bioscience and Biotechnology Research and Development, Management House, South Kalamassery, Cochin, Kerala, India; Tel: +91-484-2556315; Mob: 9847065069; Fax: +91-484-2623855; E-mail: directorsibb@scmsgroup.org

Laladhas K.P., Oommen V. Oommen and Sudhakaran P.R. (Eds.)

species *Bos indicus* and the country is blessed with a rich diversity of indigenous breeds of cow with tremendous variations from region to region. The indigenous cattle breed of Kerala, Vechur cow has gained wide public attention since the breed is facing the line of extinction as per the list of Food and Agriculture Organization (FAO). Out of the indigenous breeds of *Bos indicus* available in the country, the extreme small size, less fodder intake, easy domestication, maximum milk production compared with the body mass and high disease resistance are the beneficial features of Vechur cows. Bovine Lactoferrin (BLf) is an iron binding glycoprotein present in the milk of cows at a concentration of about 200 μg/ml and its multidimensional therapeutic properties *viz*., antimicrobial, antitumor, anti-inflammatory, antiviral, immunomodulation have been well established [1, 2]. In the present study, an attempt was made to provide molecular evidence to establish the therapeutic value of milk of Vechur cow with reference to Lactoferrin (Lf) gene.

2. LACTOFERRIN (LF) GENE

The experimental part for determining the structural and the functional features of the Lf gene includes Polymerase chain reaction (PCR) amplification of Bovine lactoferrin gene of Vechur cow (BLfV) using gene specific primers; cloning and sequencing; Single-nucleotide polymorphism (SNP) analysis of BLfV gene sequence; location of binding pockets (Fig. **3**); secondary and tertiary structure predictions; expression of the BLfV gene in eukaryotic system for the production of rBLf and the anti bacterial sensitivity test using the inoculums of bacterial strains of gram positive and gram negative bacteria for determining the bactericidal property of the rBLf protein [3-5].

As the first phase of the study the full coding region of Lf gene was amplified, cloned and sequenced. Fig. **1** demonstrates the full coding region of Lf gene amplified by gene specific primers. The sequence comparison of BLf gene of Vechur cow with the exotic breed *Bos taurus* has shown 15 SNP variations in the exon regions of the gene (Table **1**). Out of the 15 SNPs, the nucleotide polymorphism at 11 positions was associated with amino acid substitutions (Fig. **2**). The BLf gene sequences of Vechur cow are registered in National Center for Biotechnology Information (NCBI) as gene accessions: GU059864, JF926526, JF926527 and JF926528. The amino acid arginine was observed as the pronounced substitution in the Lf protein and the tertiary structure of the Lf protein confirmed the position of the arginine in β sheet, random coil and helical regions confirming its functional role (Fig. **4a** and **b**). Since arginine has vital role in natural defense mechanism as a potential immunostimulant, the higher level of arginine amino acid in Lf protein of Vechur cow milk supports the therapeutic value of the milk [6, 7]. As the common natural amino acid, the beneficial role of arginine in cell division, cell signalling, wound healing, immune function and the release of hormones even in oral supplementation has been reported by several authors. Based on the beneficial features of the Lf gene of Vechur cow observed at genomic level, the BLf gene was successfully expressed in a eukaryotic system using pOREE2 vector and CAMV35S promoter (Fig. **5**). The recombinant BLfV protein was confirmed by Western Blotting. The recombinant Lf protein of Vechur cow milk expressed in bacterial system has shown remarkable suppression of bacterial growth similar to the level of ampicillin in the strains of *E.coli*, *Staphylococcus* and *Streptococcus*, the causative pathogens of mastitis (Fig. **6a**, **b** & **c**). This is the first evidence to establish the bactericidal property of recombinant Lf of Vechur cow with reference to mastitis.

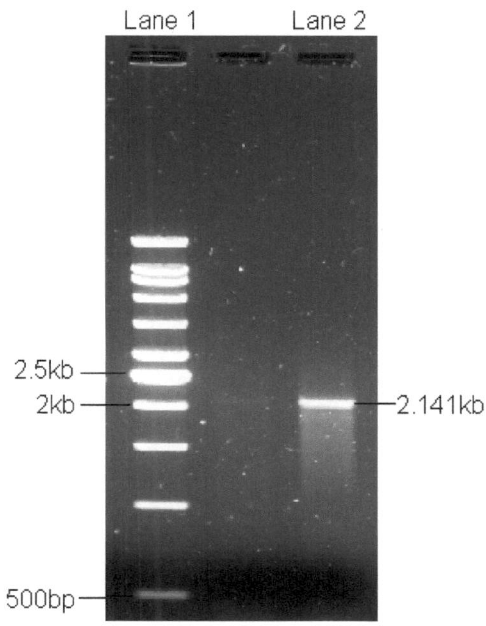

Figure 1: Bovine Lactoferrin (B Lf) gene of vechur cow.

Table 1: Nucleotide changes observed in Lf sequence of different Vechur cows and *Bos taurus*

Sample	Positions showing polymorphisms in nucleotide sequence														
Nucleotide/ Amino acid	19/7	182/61	433/ 145	526/ 176	790/ 264	979	1281	1282/428	1572	1612/538	1637/546	1786	1869/596	1879/627	1895/632
BLfBt	G/A	C/A	A/I	T/C	T/C	T	G	C/P	T	T/S	C/T	A	A/T	A/K	A/H
BLfV1	G/A	C/A	A/I	**C/R**	T/C	C	A	C/P	C	T/S	C/T	A	A/T	A/K	A/H
BLfV2	G/A	C/A	A/I	**C/R**	T/C	C	A	C/P	C	T/S	C/T	A	A/T	A/K	A/H
BLfV3	G/A	**T/V**	A/I	**C/R**	**C/R**	C	G	**T/S**	T	T/S	C/T	**T**	C/S	**G/E**	**G/R**
BLfV4	G/A	**T/V**	A/I	**C/R**	**C/R**	C	G	**T/S**	T	T/S	C/T	**T**	C/S	**G/E**	**G/R**
BLfV5	**A/T**	C/A	**G/V**	**C/R**	T/C	C	A	C/P	T	**A/T**	**A/N**	**T**	C/S	**G/E**	**G/R**

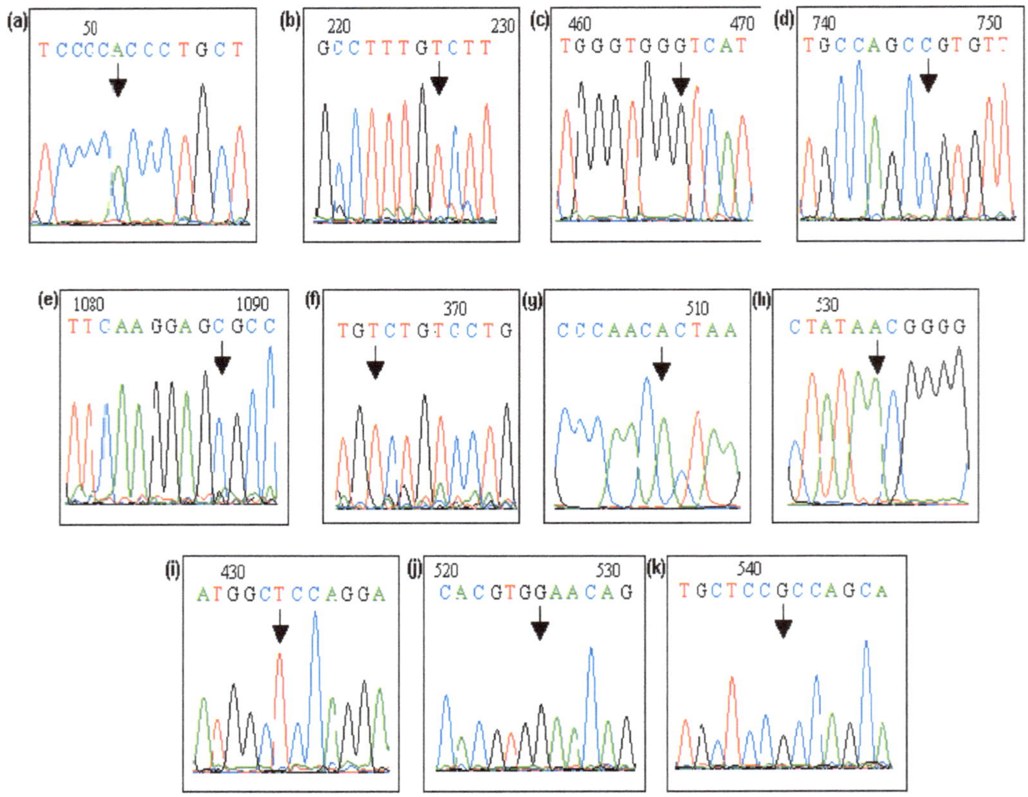

Figure 2: Chromatogram of BLf gene sequence of vechur cow showing SNPs associated with amino acid variations.

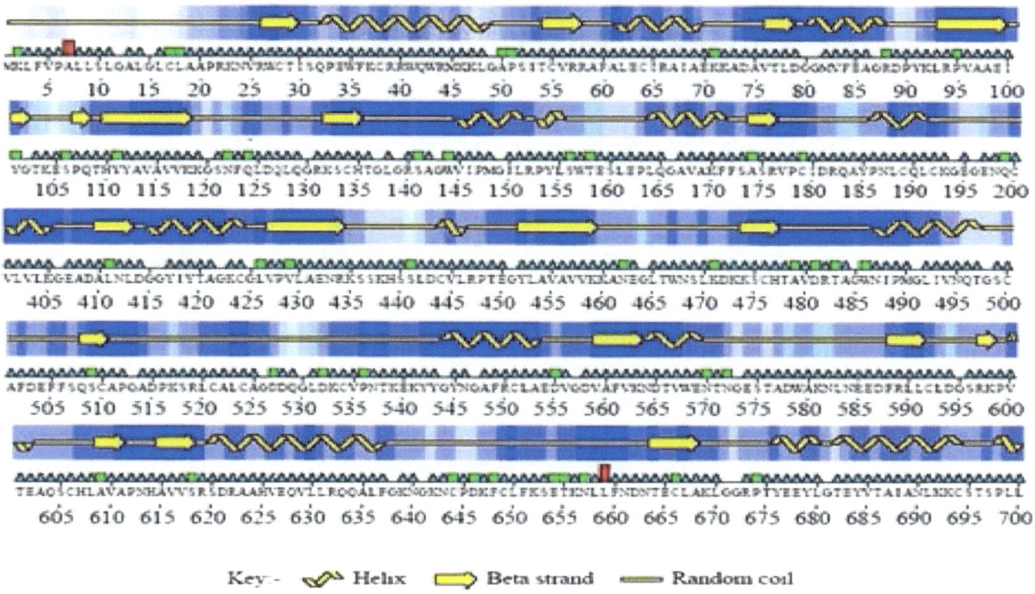

Figure 3: Binding pocket location from CASTp software.

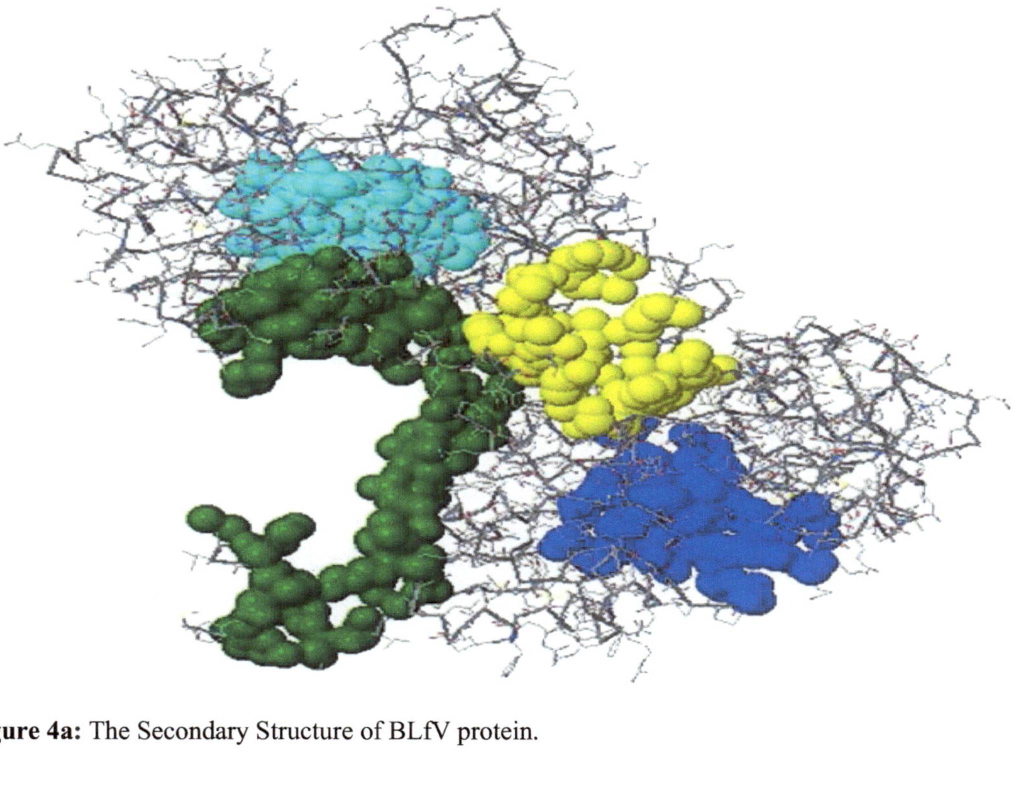

Figure 4a: The Secondary Structure of BLfV protein.

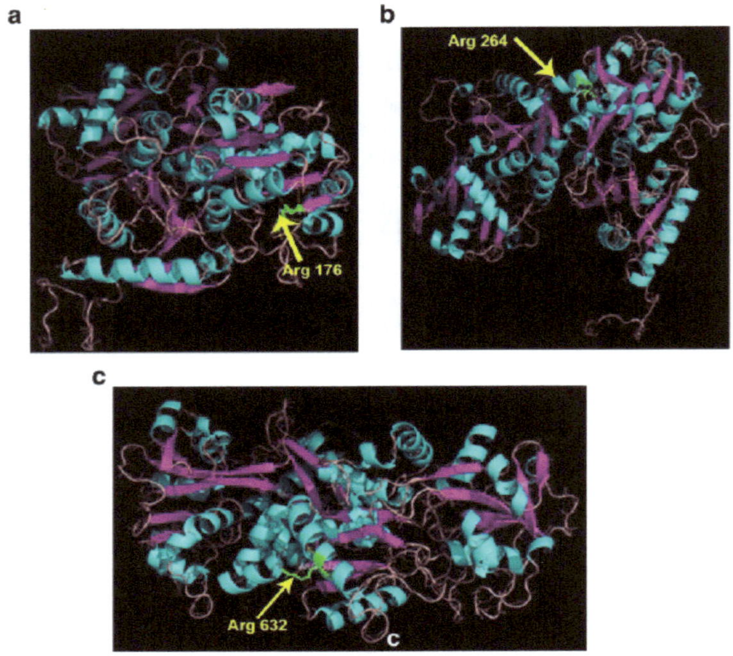

Figure 4b: Predicted Tertiary Structure of BLfV protein showing arginine substitutions.

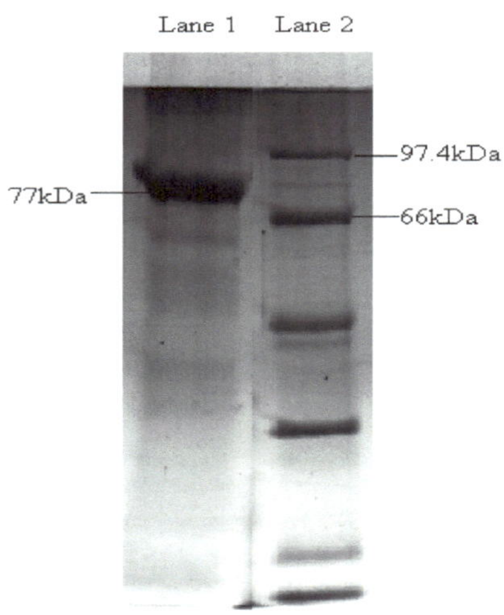

Lane 1- rBLf

Lane 2: Medium Range Protein Molecular weight marker

Figure 5: SDS-PAGE analysis of recombinant BLf protein expressed in Eukaryotic system.

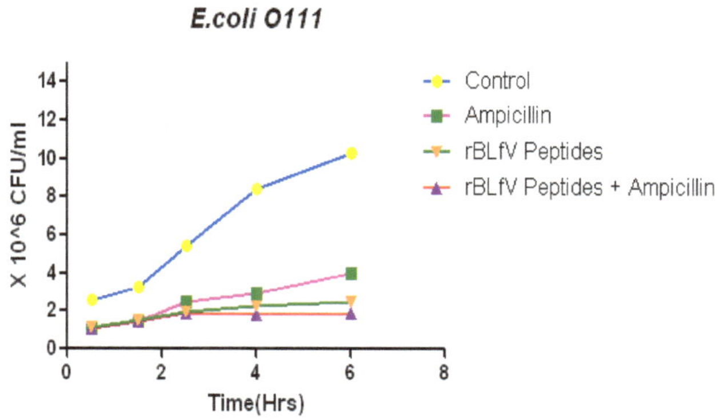

Figure 6a: Effect of rBLf peptides on the growth of *E. coli* O111.

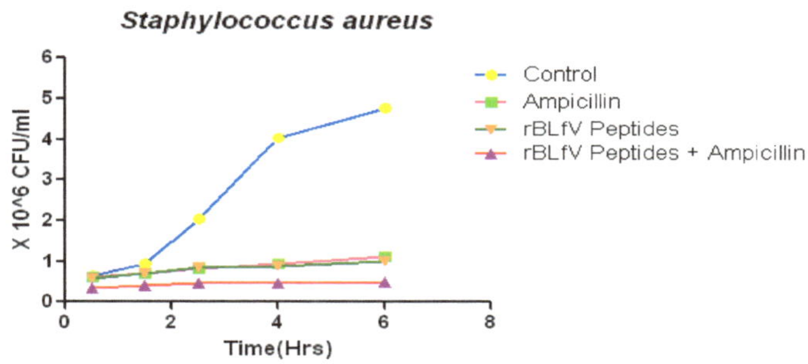

Figure 6b: Effect of rBLf peptides on the growth of *Staphylococcus aureus*.

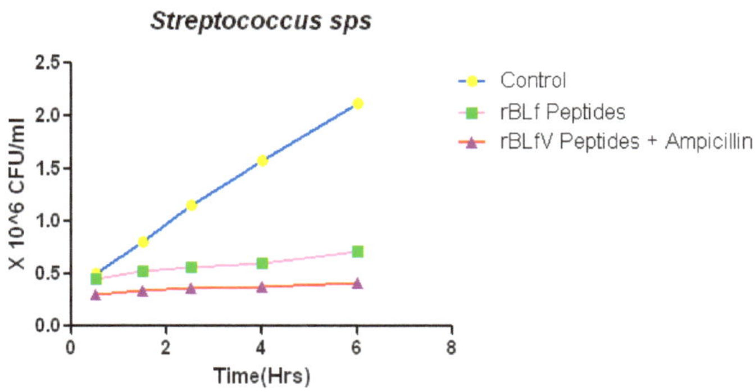

Figure 6c: Effect of rBLf peptides on the growth of *Streptococcus sps*.

3. CONSERVATION OF VECHUR COW- A NECESSITY

- Genetic erosion of domestic animal diversity has placed over a quarter of the world's breeds at risk of extinction. The preference for crossbred cows yielding more milk led to the near extinction of the native breeds of Kerala. Indigenous breeds can survive in local conditions where newer breeds would perish. Vechur cow the native breed of Kerala has low feed requirements, high disease resistance and is well adapted to the hot, humid tropical climate of Kerala, apart from being very strong and used for ploughing paddy fields. Massive cross breeding programme introduced by the government agencies as per the livestock act have suppressed the survival of the native breeds especially Vechur cow irrespective of its good qualities for domestication. Moreover lack of proper research investigations on the nutritional and therapeutic properties of the milk of Vechur breed would also be the reason for their depletion in number. It is understood that the Vechur cows show better resistance against the infectious disease mastitis but no research data exist about the probable genes responsible for the above trait. The significant factors that have to be highlighted for the conservation of a critical domestic cow are the nutritional and therapeutic relevance of its milk, the easy adaptability of the animal and its disease resistance. So the data generated from the present study demonstrate the relevance and necessity of conserving the Vechur breed as an indigenous wealth of Kerala as well as it highlights the utilization of the potential gene Lf of Vechur cow milk for exploiting its therapeutic properties in biotech industry.

ACKNOWLEDGEMENTS

The author extend their sincere thanks to Dr.G.P.C.Nayar, Chairman and Dr.Pradeep Thevanoor, the Vice-Chairman, SCMS Group, Cochin for proving the infrastructure and support for the work.

CONFLICT OF INTEREST

The authors confirm that this chapter contents have no conflict of interest.

REFERENCES

[1] O'Halloran F, Bahar B, Buckley F. Characterization of single nucleotide polymorphisms identified in the bovine lactoferrin gene sequences across a range of dairy cow breeds. Biochimie 2009; 91: 68-75.

[2] Adlerova L, Bartoskova A, Faldyna M. Lactoferrin: a review. Veterinarni Medicina 2008; 53(9): 457-468.

[3] Shashidharan A, Singh R, Bhasker S. Physicochemical characterization and functional site analysis of lactoferrin gene of Vechur cow. Bioinformation 2011; 6 (7): 275-278.

[4] Anisha S, Bhasker S, Mohankumar C. Recombinant lactoferrin (Lf) of Vechur cow, the critical breed of *Bos indicus* and the Lf gene variants. Gene 2012; 495: 23-28.

[5] Mitra A, Zhang Z. Expression of a Human Lactoferrin cDNA in Tobacco Cells Produces Antibacterial Protein(s). Plant Physiology 1994; 106: 977-981.

[6] Gokce NL. Arginine and hypertension. J Nutr 2004; 134 (10): 2818S-2819S.

[7] Stechmiller JK, Childress B, Cowan L. Arginine supplementation and wound healing. Nutr Clin. Pract 2005; 20: 52-61.

Chapter 10: An Account on Some Endemic Seeds of Southern Western Ghats

CHAPTER 10

An Account on Some Endemic Seeds of Southern Western Ghats

Anilkumar C.*, Chitra C. R., Bindu S., Maya P. V., Prajith T. M. and Joemon J.

Jawaharlal Nehru Tropical Botanic Garden and Research Institute, Palode, Thiruvananthapuram, Kerala, India

Abstract: A good collection of seeds symbolizes rich biodiversity and being viable, they signalize the continuity of life and prosperity of a nation. India, is bio-rich in terms of species holdings. In India, policies like Seed bill 2004, ascertain right over indigenous seeds. Seed listing based on their prevalent status is a basic requisite for conservation. The Seed bank of Jawaharlal Nehru Tropical Botanic Garden and Research Institute (JNTBGRI) maintains 4000 seed accessions including 1055 active; 35 base and remaining reference collections as precious futuristic germplasm holdings. Thirty seven seed samples were enlisted on the basis of their systematic and conservation status. Twenty seven species are found to be recalcitrant; eight of them orthodox while two of them are intermediate type. Though the seed recalcitrance is pre-eminent, certain specific habitat adaptive traits favour extended storage. Case study reported on *Garcinia gummi-gutta* (L.) Robson indicated the prevalence of coat imposed physical dormancy. Role of three testa layers which could be evaded by mechanical scarification for enhancing the six month 74% initial germination rate to more than 90% within two months also compliment the durability of seeds. In brief, seeds can be categorized in a way to enable stored accessions remain viable for futuristic breeding programmes.

Keywords: Dormancy, ecological adaptations, endemic, fruits, horticultural crops, recalcitrant, seed bank, survivability, Western Ghats.

1. INTRODUCTION

Indigenous seeds are playing a prominent role in breeding for new crop varieties and thus storing them for posterity is inevitable. Categorization based on seed storage requirements is a basic requisite of seed banking which was exemplified by the screening for recalcitrant characteristics of 30 agricultural and horticultural crops [1]. In this paper, an account of 37 indigenous seed samples which were

***Corresponding author Anilkumar C.:** Jawaharlal Nehru Tropical Botanic Garden and Research Institute, Palode, Thiruvananthapuram, Kerala, India; Mob: 9495729076; Tel. Off: 0472 2869226; Fax: 0472 2869646; E-mail: tsbpalode@gmail.com

Laladhas K.P., Oommen V. Oommen and Sudhakaran P.R. (Eds.)

collected during the consecutive years 2007-2011 from the southern Western Ghats is made. Seed research protocols and procedures were followed as per standardized procedures [2]. This study for comprehensive *ex situ* conservation deals with correlative effect of moisture content and viability coherent with storage temperatures. Based on the viability alterations [3, 4], 27 seeds are found to be recalcitrant, eight of them are orthodox while two of them are intermediate type. Seed types of *Calamus* are presented as a separate group since they are having more commercial importance. Usually the dormancy and recalcitrance seldom co exist except in exceptional cases as reported with *Garcinia gummi-gutta.*

2. METHODS

2.1. Collection Procedure

Mature fruits were hand harvested from the selected trees and sound seeds were sorted out to be experimented. To study the desiccation responses, fresh surface dried seed samples were kept open to laboratory conditions (60% RH and 30°C/15°C average day and night temperatures). Moisture content and the respective germination percentage were noted at an interval of every 24 hours, invariably at a reduction of 10% moisture content. Below 10% moisture content was achieved with activated silica gel and such seeds were rehydrated slowly prior to germination tests to avoid imbibition injuries. Data presented are the mean of six values ± SE.

3. RESULTS & DISCUSSION

3.1. Seed Storage

Based on the lowest feasible moisture content level with 60% viability retention, tested seeds were categorized into eight orthodox, twenty seven recalcitrant and two intermediate types. Correlation of species distribution with their respective seeds indicates omnipresence of all categories irrespective of gradation of high (above 700); mid (300- 700) and low (50-300 msl) altitudes. Nevertheless, the seed desiccation responses are more suitable to their microclimatic ecology such as open hillocks, riparian, misty areas *etc*. This is supported by the response of orthodox seeds of *Rauvolfia micrantha* and recalcitrant *Semecarpus auriculata* seeds of mid altitudinal distribution (Table **1**).

Within the genus *Calamus*, [5], there is little taxonomic relationship among species exhibiting recalcitrance. Among the listed eight endemic *Calamus* species, five are recalcitrant; two orthodox and *C. travancoricus* is an intermediate type. This commercially important genus which is an ecological indicator of a viable biome, propagates solely through seeds. Among the majority of the readily germinable recalcitrant species, *C. travancoricus* seeds exhibited dormancy with comparatively low initial moisture content value of 24%. The viability which was retained up to 10% moisture content was abruptly lost on exposure to 10°C, as typical of intermediate type (Table **2**).

Table 1: Categorisation of 29 endemic seeds of southern Western Ghats

Species name	Family, Altitude of distribution in meter and Conservation Status	Fresh Moisture content (%) & germination %	Lowest feasible Moisture content (%) & germination %	Category
1.*Buchnania barberi* Gamble	Anacardiaceae Upto 200 Critically endangered	22.42 ± 0.03, 84 ± 1.87	9.6 ± 0.5, 78 ± 2.5	Orthodox
2.*Decalepis hamiltonii* Wight & Arn.	Periplocaceae 300-1200,Endangered	11.5 ± 0.2, 80 ± 1.8	4.9 ± 0.18, 70 ± 1.5	Orthodox
3.*Ensete superbum* (L.) Cheesm.	Musaceae Upto 100 Rare & endangered	17.5 ± 0.28, 30.8 ± 1.5 Dormant	2.5 ± 0.08, 51± 1.7	Orthodox
4.*Calophyllum apetalum* Willd.	Clusiaceae Upto 900,Vulnerable	22.1 ± 0.5, 83 ± 2.5	4.9 ± 0.18, 60 ± 1.5	Orthodox
5.*Rauvolfia micrantha* Hook. f.	Apocynaceae 100-600 Endangered	19.7 ± 0.15, 62 ± 0.5 Dormant	4.8 ± 0.25, 50 ± 1.6	Orthodox
6.*Rauvolfia hookeri* Srinivas *et* Chithra	Apocynaceae 600-1500,Endangered	24.5± 0.25, 65± 2.5 Dormant	4.9 ± 0.01, 75± 1.8	Orthodox
7. *Janakia arayalpathra* (Joseph & Chanthras.) Venter (*Decalepis arayalpathra* (Joseph & Chanthras.) Venter)	Periplocaceae 800-1200 Critically endangered	36 ± 0.35, 40 ± 1.6 Seed setting problem	31± 1.1, 21.9 ± 1.2	Orthodox
8.*Thottea duchartrei* Sivarajan, Babu & Indu Balachandran	Aristolochiaceae Upto 800,Vulnerable	18 ± 0.5, 53 ± 2.5	8 ± 0.15, 50 ± 1.5	Orthodox
9. *Actinodaphne bourdillonii* Balakr.	Lauraceae,1000-1800	18.3 ± 0.35, 95 ± 2.7	5± 1.01 61 ± 1.9	Intermediate
10.*Ixora brachiata*	Rubiaceae,300-1800	51 ± 1.8,	9.6 ± 0.5,	Recalcitrant

Table 1: contd…

Roxb.ex.DC		96 ± 1.1	49 ± 2.9	
11.Salacia malabarica Gamble	Hippocrateaceae Upto 300,Endangered	49.2 ± 1.02, 95 ± 2.7	30 ± 1.5, 50 ± 1.58	Recalcitrant
12.Cullenia exarillata Robyns	Bombacaceae,700-1300	50 ± 0.71, 96.6 ± 1.02	30 ±1.5, 70 ± 0.94	Recalcitrant
13.Goniothalamus rhynchantherus Dunn	Annonaceae,600-1200 Rare	21.9 ± 0.3, 70.2 ± 1.5	9.7 ± 0.67, 0	Recalcitrant
14.Piper barberi Gamble	Piperaceae,600-1000 Rare & Critically endangered	33.21 ± 0.7, 96 + 2.9	21.29 + 0.5, 24.5 + 2.5	Recalcitrant
15.Popowia beddomeana H.K.f.& T.Bedd.	Annonaceae,600-1200 Endangered	32.03 ± 0.4, 31.8 ± 0.96 Seed setting problem.	27.6 ± 1.3, 0	Recalcitrant
16.Garcinia gummi-gutta (L.) Robson	Clusiaceae, Upto 1200 Vulnerable	42.1 ± 0.87, 51± 0.7 Dormant	33.2 ± 0.8, 50 ± 1.5	Recalcitrant
17.Syzygium travancoricum Gamble	Myrtaceae,300-1200 Critically endangered	53.2 ± 0.9, 70 ± 1.58 Seed infection	45 ± 1.6, 40 ± 0.9	Recalcitrant
18.Humboldtia decurrens Bedd. ex Oliver.	Caesalpiniaceae, Upto 1200,Vulnerable	41.3 ± 0.96, 88 ± 0.9	30 ± 1.5, 63 ± 1.3	Recalcitrant
19.Myristica malabarica Lam.	Myristicaceae,300-1200,Vulnerable	27.7 ± 0.5, 97 ± 0.9	11.01 ± 0.1, 0	Recalcitrant
20. Ormosia travancorica Bedd.	Fabaceae,1200-1800 Endangered	24.8 ± 0.8, 83.5 ± 1.5	16 ± 0.45, 50.4 ± 1.3	Recalcitrant
21.Ochlandra wightii (Munro) Fischer	Poaceae, Upto 1000	53.7± 0.6, 96 ± 1.8	42.8 ± 1.01, 58 ± 2.5	Recalcitrant
22.Pinanga dicksonii (Roxb.) Blume	Arecaceae,1200-1800	22.8 ± 1.01, 88 ± 2.5	15 ± 0.03, 69 ± 1.87	Recalcitrant
23.Garcinia indica (Thouars)Choisy	Clusiaceae,300-1800 Vulnerable	53.4 ± 0.92, 97 ± 2.1	19.7± 0.7, 60.2 ± 1.6	Recalcitrant
24.Semecarpus auriculata Bedd.	Anacardiaceae,300-600,Threatened	32.7 ± 0.7, 73± 2.5	21.2 ± 0.49, 31 ± 1.9	Recalcitrant
25.Arenga wightii Griffith	Arecaceae,600-1400 Threatened	29 ± 0.5, 40.6 ± 1.7,Seed setting problem	19.8 ± 1.31, 0	Recalcitrant
26.Hydnocarpus macrocarpa (Bedd.)	Flacourtiaceae, Upto 400,Vulnerable	52.8 ± 1.01,	36 ± 1.18,	Recalcitrant

Table 1: contd…

Warb.		59.5 ± 1.69	50.2 ± 1.59	
27.Garcinia imberti Bourd	Clusiaceae,600-1200	50 ± 1.58, 80 ± 0.97	29.4 ± 1.69, 31.5 ± 1.01	Recalcitrant
28.Hopea parviflora Bedd.	Dipterocarpaceae,300-1200,Endangered	41± 1.7, 95.6 ± 1.7	30.6 ± 0.6, 50.8 ± 1.5	Recalcitrant
29.Vateria indica L.	Dipterocarpaceae,300-1200,Critically endangered	43.9± 1.7, 90 ± 3.5	31.5 ± 1.71, 50 ± 2.3	Recalcitrant

All data are mean of six values ± SE

Table 2: Categorisation of eight endemic *Calamus* seeds of southern Western Ghats

1.Calamus brandisii Becc	Arecaceae,1000-1500	30.6 ± 0.13, 95 ± 0.05	10 ± 0.13, 88 ± 0.13	Orthodox
2.Calamus gamblei Becc.	Arecaceae,700-2000	34.4 ± 0.13, 98 ± 0.2	11.5 ± 1.5, 51 ± 0.5	Recalcitrant
3Calamus hookerianus Becc.	Arecaceae,50-1000	19.2 ± 0.14, 97 ± 2	5.1 ± 0.19, 70 ± 1.5	Orthodox
4.Calamus karnatakensis Renuka & Lakshmana	Arecaceae,500-1200	25 ± 1.7, 97 ± 2	16 ±0.9, 50 ± 1.3	Recalcitrant
5.Calamus metzianus Schlecht	Arecaceae,About 50	27.6 ± 1.1, 97 ± 0.25	11.6 ± 0.97, 49 ± 1.7	Recalcitrant
6.Calamus shendurunii Anto,Renuka and Sree Kumar	Arecaceae,400-1000 Endangered	33.08 ± 0.4, 98 ± 2	25.1 ± 1.7, 50 ± 1.9	Recalcitrant
8. Calamus travancoricus Bedd.ex Becc.& Hook	Arecaceae,300-500	24.24 ± 0.09, 52.02 ± 2.5 Dormancy	10.3 ± 0.7, 50 ± 1.6	Intermediate
9.Calamus vattayila Renuka	Arecaceae,250-1000 Endangered.	24.2 ± 0.37, 97± 2.7	15.3 ± 0.9, 40 ± 1.7	Recalcitrant

All data are mean of six values ± SE

Janakia arayalpathra (*Decalepis arayalpathra*), *Popowia beddomeana*, *Arenga wightii* exhibit seed set problems. Purity percentage of *Janakia arayalpathra* was very low since most of them were chaffy. Seeds of *Popowia beddomeana* and *Arenga wightii* are recalcitrant seeds with initial 30% moisture content but still with only low initial germination percentage of 30-40 range. Seeds of *Calamus travancoricus, Ensete superbum, Garcinia gummi-gutta, Rauvolfia micrantha* and *Rauvolfia hookeri* show dormancy. Dormancy is regarded as the failure of an intact viable seed to complete germination under favourable conditions [6]. Dormancy is an advanced adaptive feature for ensuring species survivability as in the case of orthodox seeds while recalcitrant seeds are considered as primitive and they usually germinate readily. There are exceptions as in the case of recalcitrant *Garcinia gummi-gutta* seeds which remain dormant and germinate sporadically. Dormancy of *Garcinia gummi-gutta* is due to seed coat as it impedes water entry.

At the sloughy forest floor, seeds are slowly decoated and germinated after a gap of approximately 5 to 6 months. Since these seeds cannot be dried below 20% from the initial 41.4% moisture content, such recalcitrant seeds retain viability only upto two months. Detailed investigation of seed coat revealed its three layered nature; each seed coat layer has its own role in controlling dormancy. The physical dormancy due to seed coat relays on the viability retention by six month open stored seeds (28±2°C, 60%RH) (Table **3**).

Table 3: Role of seed coat layers on *Garcinia gummi-gutta* dormancy and durability

Treatments	**Percentage germination**					
	Ist month	**IInd month**	**IIIrd month**	**IVth month**	**Vth month**	**VIth month**
3rd layer (Control)	0 100	0 100	0 95±2.2	6±8.9 91. ±3.1	28±8.3 76±2.1	78±13.03 71.7±3.1
IInd layer	0 100	0 98.3±1.7	36±8.94 96.7±2.1	56±8.94 80±2.6	96±5.5 61.7±3.1	- 45±2.2
1st layer	12±4.5 100	16±5.5 98.3±1.7	74±8.9 76±2.1	90±7.1 61.7±3.1	96±5.5 43±3.3	- 0
De-coated seeds	74±8.9 96.7±2.1	98±4.5 91. ±3.1	100 46.7±2.1	- 38.3±1.6		

All germination data are mean of six values ± SE. Value in italics denote viability on exposure to 28±2°C, 60%RH.

CONCLUSION

Seed as an entity is a living model for an enigmatic future biome wherein climate consistently changes. By recording the prevailing features of some selected seeds, remodelling of species regeneration protocols becomes possible since previous adaptations may provide clues for further efforts on conservancy. This chapter is primarily an account of the viability of 37 endemic seeds in response to their storage water physiology. Species status, genus wise grouping and ecological adaptations with respect to the southern Western Ghats are considered to be the base data on seed biology. It becomes imperative that species survival among altitudinal gradations often share more adaptive features irrespective of their varying fruit/seed developmental traits. Studies on seed behaviour of different species of the fringe areas of same biomes, as noted in the case of *Rauvolfia micrantha* and *Semecarpus auriculata* may augment species restoration. Case study on dormancy is selected due to its relevance on seed durability and the fact that all recalcitrant seeds are not readily germinable as in the case of *Garcinia gummi-gutta.* This addresses concerns on nursery practices. *Guttifera* type germination as a successfully evolved seed trait favour species survival that surpasses predation wherein plants and animals co exists in a biome.

ACKNOWLEDGEMENTS

Authors acknowledge Director, JNTBGRI for facilities and Kerala Forest Department for financial assistance.

CONFLICT OF INTEREST

The authors confirm that this chapter contents have no conflict of interest.

REFERENCES

[1] Chin HF, Hor YL, Mohdd Lassim MB. Identification of recalcitrant seeds. Seed Sci. & Technol 1984; 112: 429-436.

[2] ISTA. ISTA ordinary meeting 2005. Technical committee session report. Seed Testing International ISTA News Bulletin 2005; 130: 33-41.

[3] Roberts EH. Predicting the storage life of seeds. Seed. Sci. and Technol 1973; 1: 499-514.

[4] Ellis RH, Hong TD, Roberts EH. An intermediate category of seed storage behavior? I Coffee.J. Exp. Bot 1990; 41: 1167-1174.

[5] Berjak Patricia, Pammenter NW. From Avicennia to Zizania: Seed Recalcitrance in Perspective (Review). Annals of Botany 2008; 101: 213-228.

[6] Bewley JD. Seed germination and dormancy. The Plant Cell 1997; 9: 1055-1066.

Chapter 11:
Non Wood Forest Biodiversity for Food Security Rattans a Case Study

CHAPTER 11

Non Wood Forest Biodiversity for Food Security-Rattans a Case Study

Manohara T. N.*

Rain Forest Research Institute, Jorhat, Assam, India

Abstract: Studies were conducted on two rattan species of the North East India, *Calamus flagellum* Griff. ex Mart. and *C. floribundus* Griff. (Arecaceae), to know the nutrient value of the shoots. In general, the rattan species studied contain considerable amounts of protein, carbohydrate, fibres, vitamins and minerals. When compared to other palm shoots and bamboo shoots the level of sodium and fat are very low. The level of Vitamin B complex was at par with other leafy vegetables and bamboo species. Phosphorous content was also more in rattans than in the leafy vegetables. Nutritional features of rattan shoots make them an excellent source of dietary supplements to local people. The rattans can be promoted among farmers as under storey crops in an agroforestry system and in shifting cultivation to provide enhanced livelihood options. Large scale cultivation of edible Rattans with community participation will reduce the pressures on harvesting from the wild and promote conservation of the rich diversity of Rattans in India.

Keywords: Agroforestry, *Calamus*, dietary fibres, edible shoots, lowering cholesterol, niacin, nutrient, Rattans, thiamine.

1. INTRODUCTION

Rattans are considered as a valuable forest resource of considerable socio-economic value. Globally, there are about 600 rattan species belonging to 13 genera; 51 species belonging to 4 genera occur in India. Rattans are mainly confined to three geographic regions in India - the Western Ghats, the Andaman Nicobar Islands, and Northeast India. Over exploitation and habitat destruction in recent years have led to a rapid decline in the natural populations of rattans. Many of the species are on the verge of extinction [1-4]. Several species of rattans supply edible shoots. The shoots can be used in preparation of delicious food and nutritious dishes. Rattan shoot is a traditional dish in some Asian countries and is considered as a delicacy in Europe and America. Rattan cultivation for its edible shoots is in practice in many South-East Asian Countries. In India, in the North-Eastern States and Western Ghats it is used as food as well as medicine (Fig. **1**).

***Corresponding author Manohara T. N.:** Rain Forest Research Institute, Jorhat, Assam, India; My Tel. (Off): 03762305168; Mobile: +91 9435351304; E-mail: manohara_tn@yahoo.com

Laladhas K.P., Oommen V. Oommen and Sudhakaran P.R. (Eds.)

Figure 1: Tribal woman selling rattan shoots in a forest village near Dehing-Patkai WLS, Assam, India. (Note: Arrow mark showing rattan shoots).

2. EDIBLE RATTAN SPECIES

Two edible rattan species from North East India, *Calamus flagellum* (raidang) and *C. floribundus* (Lesai beth or bath beth) were selected based on the data available from the published literature, field survey and local knowledge from indigenous people/tribal's. Shoot samples were selected from the healthy plants growing in Gibbon Wild Life Sanctuary, Assam. The top two to three feet stem was cut and the outer spiny leaf sheaths were removed and inner soft core portion were selected for nutrient analyses.

3. NUTRITIONAL CONTENT OF RATTANS

The samples were analyzed as per the standard protocols [5, 6]. The analyses were done partly at the Rain Forest Research Institute laboratory, Jorhat and at the Federation of Chambers of Commerce and Industry (FICCI), Research and Analyses Centre, New Delhi. A comparative study of the nutrient contents was made with other well accepted edible rattans, palms, leafy vegetable and bamboos (Table **1**).

The analysis of data (Table **1**) shows that the rattan shoots are rich in proteins, carbohydrates, and contains considerable amount of calories in addition to Vitamin B complex, mineral elements and dietary fibres. The level of sodium is very low. The level of Vitamin B complex was at par with other leafy vegetables and bamboo species. Of particular interest is the presence of higher levels of vitamin B_3 (Niacin) in *C. flagellum* and *C. floribudus* (17.42 mg/100g and 18.30 mg/100g, respectively), the highest for the vegetables reported so far. Vitamin B_3 or Niacin (daily requirement for adult is 16-18 mg) plays a vital role in fatty acid metabolism, in lowering cholesterol levels, regulating blood sugar levels and in controlling the blood pressure. Phosphorous content was also more in rattans than in the leafy vegetables. The presence of high percentage of protein from two species of Chinese

Table 1: Comparison of nutrient values of *Calamus* with other palm genera, leafy vegetables and bamboo s

Parameters	Samples								
	C. flagellum	*C. floribundus*	*C. tenuis*	*C. rivalis*	*C. thwaitessi*	*C. hookerianus*	*Cocos nucifera*	*Borassus flabellifer*	*Arenga pinnata*	Amaranth
Moisture %	89.90	89.87	86.30	88.50	91.30	59.60	~	~	~	~
Crude Protein % by wt	09.75*	17.93*	15.00	14.00	18.13	11.88	13.2000	8.9000	1.9000	4.9000
Fat % by wt.	Not Detected	0.12	~	~	~	~	~	~	~	~
Carbohydrate % by wt.	7.50	5.59	~	~	~	~	~	~	~	~
Energy, Kcal/100g	36.96	35.32	~	~	~	~	~	~	~	~
Sugar as Sucrose % by wt.	Not Detected	Not Detected	~	~	~	~	~	~	~	~
Vitamin C mg/100g	Not Detected	Not Detected	~	~	~	~	~	~	~	~
Crude Fiber % by wt.	27.32	14.09	1.90	2.00	18.13	11.88	7.4000	7.2000	9.4000	~
Phosphorous %	34.87	51.52	0.27	0.30	0.34	0.34	0.4400	0.4600	0.0560	0.1000
Total Ash %	0.96	1.45	~	~	~	~	~	~	~	~
Vitamin B mg/100g	Not Detected	Not Detected	~	~	~	~	~	~	~	~
Vitamin B_1 mg/100g	0.16	1.19	~	~	~	~	~	~	~	~
Vitamin B_2 mg/100g	0.0095	0.26	~	~	~	~	~	~	~	~
Vitamin B_3 mg/100g	17.42	18.30	~	~	~	~	~	~	~	~
Vitamin B_{12} mcg/100g	Not Detected	Not Detected	~	~	~	~	~	~	~	~
Iron, mg/100g	5.6	6.39	0.18ppm	1.03ppm	0.32 ppm	0.55ppm	0.0037	~	0.0090	0.0210

Table 1: contd….

Zinc mg/100g	1.07	1.56	0.17ppm	0.22ppm	0.71ppm	0.59ppm	~	~	~	
Magnesium mg/100g	4.57	37.45	6.22ppm	9.58ppm	5.64ppm	13.62ppm	~	~	~	
Selenium mcg/100g	0.5	0.5	~	~	~	~	~	~	~	
Calcium mg/100g	18.42	61.04	6.88ppm	8.92ppm	13.68ppm	7.39ppm	0.2000	0.6000	0.4000	0.
Potassium mg/100g	400.90	363.29	~	~	~	~	~	~	~	
Sodium mg/100g	6.22	4.11	~	~	~	~	~	~	~	
Fat Soluble										
Vitamin A, IU/100g	Not Detected	Not Detected	~	~	~	~	~	~	~	
Vitamin D, IU/100g	Not Detected	Not Detected	~	~	~	~	~	~	~	
Vitamin E IU/100g	Not Detected	Not Detected	~	~	~	~	~	~	~	
Vitamin K IU/100g	Not Detected	Not Detected	~	~	~	~	~	~	~	

✓ Data taken from Wealth of India [7], Renuka *et al.,* [8], Chongtham *et al.,* [9], and USDA National Nutrient Database [10].
*Converted to crude protein (Enriched protein value x 6.25 = crude protein); ~ = data not available

of Chinese rattans, *Daemonorops margaritae* (Hance) Becc. and *Calamus simplicifolius* C.F. Wei [11] has been reported. They found that total proteins in fresh shoots of *C. simplicifolius* is higher than the average content of total proteins in 12 kinds of vegetables (1.45%) and 27 kinds of bamboo shoots (2.65%). The protein content of *D. margaritae* was higher than the average protein content of vegetables and lower than the average in bamboo shoots. A similar observation was made for Indian and Thai rattans [12]. Higher percentages of calcium and zinc were also noticed in these two species. The fibre content increases with age when other nutrient contents decrease. Three to six months old shoots are best suited for consumption.

4. SOCIO ECONOMIC VALUE AND CONSERVATION OPTIONS

India host a great diversity of rattans, but because of unsustainable collection from wild habitats, deforestation and land use changes the gene pool diversity of rattans is under threat. Further in depth research on the nutritional value of diverse rattan species will help in ensuring food security to rural poor.

Rattan shoot farming has immense potentialities in the rural sector for raising additional income. Rattan shoots used for consumption in India are, at present, mostly extracted from the wild. As most of the species are facing severe declines in their natural populations, the cultivation of rattans on farm lands and shifting cultivation lands not only aids in their conservation but also generates additional sources of income to the rural populace. India with its vast rattan resources has considerable scope for cultivation of edible rattan species in several areas of the country. In Northeast India in many places where shifting cultivation still prevails rattan cultivation takes on greater significance. It can be promoted in agroforestry systems, shifting cultivation (Jhum) fields and in ecorestoration programmes, thereby reducing pressure on natural populations in forest. There is scope for producing more value added products and technological improvement for long term preservation of rattan shoot. Large scale cultivation of edible Rattans with community participation will reduce the pressures on harvesting from the wild and promote conservation of the rich diversity of Rattans in India. They need to be popularized as food and commercial crops. If their cultivation is taken up on a large scale rattans could contribute significantly to the rural economies and nutritional need of rural populace.

ACKNOWLEDGEMENTS

The financial support from Indian Council of Forestry Research and Education (ICFRE), Dehradun is gratefully acknowledged. The author is thankful to Director, RFRI for whole hearted support and encouragement.

CONFLICT OF INTEREST

The author confirms that this chapter contents have no conflict of interest.

REFERENCES

[1] Basu SK.Threatened palms of India - Some case studies. J Econ Taxon Bot 1986; 7: 493-497.

[2] Johnson D, Ed. and the IUCN/SSC Palm Specialist Group. Palms: Their conservation and sustained utilization. Status Survey and Conservation Action Plan. Gland, Switzerland and Cambridge, U.K.: IUCN 1996.

[3] Renuka C. Rattans of North eastern India - A cause for great concern. Arunachal Pradesh Forest News 1996; 14(2): 8-11.

[4] Sastry C. Rattan in the twenty first century: an outlook. In: Dransfield J, Tesoro FO, Manokaran N, Eds. Rattan: Current Research Issues and Progress for Conservation and sustainable Development. Rome, Food and Agriculture Organization of the United Nations (FAO) 2002; pp. 9-22.

[5] AOAC. Official methods of analysis, 18th edition. Dr. William Horwitz and Dr. George W. Latimer, Jr., Eds., Maryland, USA: AOAC International, (Revision–1, 2006) Chapter 4.65.

[6] APHA. American Public Health Association. Standard Methods for the Examination of Water and Wastewater, 18th Edition, 1992; Test method: 3030 D-J.

[7] Wealth of India - An Encyclopedia of India's Raw Material Resources, 1948-1976. New Delhi, India: National Institute of Science, Communication and Information Resource (CSIR).

[8] Renuka C, Thomas JP, Rugmini P. Cultivation of rattan for edible shoots - A source of additional income for the farmers of the Western Ghats. Kerala Forest Research Institute Research Report 2007; 1127: 49-59.

[9] Chongtham N, Bisht MS, Haorongbam S. Nutritional properties of bamboo shoots: Potential and prospects for utilization as a health food. Food Science and Food Safety 2011; 10(3): 153-168.

[10] USDA. USDA National Nutrient Database for Standard Reference, Release 25 available from http://www.ars.usda.gov/services/docs.htm?docid=8964, 2012. (28 May 2013).

[11] Xu HC, Zhou ZZ, GT Yin. Nutritional evaluation of shoots of two rattan species. RIC Bulletin 1991; 10(1): 17-20.

[12] Renuka C, Thomas JP. An evaluation of the nutrient contents of the edible shoots of four species of rattans from India. J. Non-Timber Forest Products 2006; 13(3):173-177.

CHAPTER 12

Collective Action on Improving Environmental and Economic Performance of Vegetable Production: Exploring Pesticides Safety in India

Sreejith Aravindakshan[1,*] and Sherief A. K.[2]

[1]*Farming Systems Ecology, Wageningen UR, the Netherlands and* [2]*Department of Extension, Kerala Agricultural University, India*

Abstract: From the chemical input-intensive yield-enhancement practices of the Green Revolution era, agricultural research and development focus is gradually shifting towards establishing Good Agricultural Practices (GAP) in fruits and vegetable sector. The dominant problems affecting fruits and vegetables in terms of safety is presence of pesticide residues. Globally, authorities have long highlighted this risk and imposed appropriate maximal limits of residues (MLRs). In spite of imposing MLRs in fresh vegetables, negative health effects of pesticides residue in consumers have been increasingly reported from states of India like Kerala. Along with other factors, food quality and safety declination resulting from inappropriate chemicals and pesticide use during crop production is widely documented as one of the root-causes of the health issues. The weak quality assurance schemes in developing countries impede smallholders' inclusion in high value chains due to imperfect institutional and governance arrangements throughout the system. Apparently, erstwhile studies have emphasized the significance of collective action among smallholders as a solution to the above constraints. However, past studies on vegetable production in India are either from a horticultural and entomological perspective on increasing production and productivity, pest management or on cost of cultivation and those solely from an econometric, institutional and collective action perspective have been hardly studied. Still lesser are studies understanding the inter-linkages between smallholders' collective action and pesticides risk reduction in vegetable production of India. In this backdrop, the current study examines various econometric models and suggests suitable models to assess the institutional mechanisms on improving environmental and economic performance of vegetable production in India under collective action.

Keywords: Agricultural practices, bio-control agents, crop rotation, data envelopment, ecological conditions, food safety, insecticides, organic farming, pathogenic, pesticide safety, pest management, poverty reduction, quality assurance, smallholders, vegetable production.

***Corresponding author Sreejith Aravindakshan:** Farming Systems Ecology, Wageningen UR, the Netherlands; Tel: +31- 633 472670; Fax: +31- 317 418094; E-mail: sreejith.aravindakshan@wur.nl

Laladhas K.P., Oommen V. Oommen and Sudhakaran P.R. (Eds.)

1. INTRODUCTION

From the chemical input-intensive yield-enhancement practices of the Green Revolution era, agricultural research and development focus are gradually shifting to establishing Good Agricultural Practices (GAP) in fruits and vegetable sector. The dominant problems affecting fruits and vegetables in terms of safety do not concern pathogenic contamination but pesticide residues [1]. Globally, authorities have long highlighted this risk and imposed appropriate maximal limits of residues (MLRs) [1, 2], and promoted integrated pest management (IPM) [3] and organic farming practices. In spite of imposing MLRs in fresh fruits and vegetables, negative health effects of pesticides residue in consumers, have been increasingly reported [4] (for example recent cases of endosulphan victims in Northern Kerala). Along with other factors, food quality and safety declination resulting from inappropriate chemicals and pesticide use during crop production is widely documented as one of the root-causes of the health issues [5-8].

The role of both public agencies and private actors in pesticide risk reduction have been widely discussed in the academic literature [9, 10]. In fact, there is a common consensus between the private and public food safety regulations in developed countries, where both these regulatory standards increasingly wreathe together to manage and ensure food safety. On the other hand, the institutional frameworks on food safety regulations in developing countries are significantly weaker, and controlled by government bodies [10, 11]. Besides, with private voluntary standards still in the state of emergence, there is a lack of clarity in the structure and functioning of food safety regulations in developing countries [3].Such weak quality assurance schemes impede smallholders' inclusion in high value chains due to imperfect institutional and governance arrangements throughout the system. Apparently, erstwhile studies have emphasized the significance of collective action among smallholders as a solution to the above constraints [12, 13]. However, past studies on fruits and vegetable production in South Asian countries like India are either from an agronomic perspective on increasing production and productivity [14, 15] or on cost of cultivation [16] and those solely from an econometric, institutional and collective action perspective have been hardly studied. Still, the studies about understanding the inter-linkages between smallholders' collective action and pesticides risk reduction in vegetable production of South Asia are few.

1.1. Pesticide Safety in Indian Scenario

In South Asia, with an annual production of approximately 81 million tonnes of fruits and 162 million tonnes of vegetables during 2012-2013; India is regarded as the leading horticultural producer of the world [17]. The country has witnessed excessive pesticide use in case of both fruits and vegetables [18, 19]. Because of varying socio-economic conditions and lack of knowledge, farmers in India tend to misuse or overuse pesticides that are illegal, banned, and acutely toxic or in most cases not fit for horticultural crops [20].

Inspite of India's status as the second largest producer of fresh fruits and vegetables, export of horticultural products is less [21]. For a long time, India has aspired to export its fruits and vegetables to European markets. This was in accordance to the country's rural development strategy aimed to promote exports of agricultural goods [22]. Since early 1990's, India has achieved short term success in export of horticultural products to Northern European countries like the UK, Netherlands and Germany [23]. Apparently, there are success stories in fruits and vegetables export involving cooperatives and collective action. For example: grapes' export to UK, Germany and Netherlands by *'Mahagrapes'* in the Indian state of Maharashtra [21], and vegetable farming by *'Kudumbashree'* women collective groups in Kerala state [24].

Moreover, it is often reported that smallholders limited by resources may not be able to cope up with the safety standards of the global and domestic markets [13, 25]. Inability to comply with the various standards acts as an exclusion barrier for smallholder production systems to enter high value chains [3]. As such there is an on-going debate around the rise of food safety standards, especially private agri-food standards – whether these act as a barrier or benefit to the smallholder producer at the downstream of the agri-food chains in developing countries like India and impede poverty alleviation [26, 27].

2. ENSURING PESTICIDE SAFETY FOR HIGH VALUE CHAINS

Under the realms of food safety in pest management, actions by individual farmers alone in case of vegetable production could cause new problems. One of the often observed problems in vegetable production of India is the extensive use of pesticides on individual farms neglecting pesticide translocation effects to nearby plots, water resources or soil. This could also trigger pest movement to fields not treated with pesticides, or pests developing localized resistance. Another problem may arise when predator population is below viable size in fields that uses biological means of control. Furthermore, in organic vegetable production, there are other issues with respect to pest management. Primarily, translocation effect of pesticides from non-organic farms could damage the positive effects of organic farming and secondarily, there is the issue of non-uniformity in practices between the organic farms.

Collective action in agriculture is identified as providing several benefits to small and marginal farmers such as reducing transaction costs while increasing bargaining opportunities, alongside better access to inputs, credit, and market [28-30].

Most importantly, the impact of collective action in managing pesticide safety standards occur when farmers in a particular area or village take decisions jointly and share operations and costs to implement integrated pest management practices. For *e.g.* joint decisions on, and the collective use of organic management techniques, bio-control agents, inorganic less toxic remedial measures, or the use of recommended crop and pest specific pesticides in combination with sequential/companion cropping with "*pest push and pull techniques*", as per requirement.

Forming farmer collectives in pest management for participation in high value chains could be challenging in developing countries due to farmers' heterogeneity in social and economic status, belief systems, attitudes, behavior and values towards the society and the environment *etc*. Farmers behave rationally and often try their best to produce more from limited land, and often venture into extensive pesticide application or compete with neighboring farmers for markets and better prices. Therefore, it is important to convince all the farmers in an area to adopt integrated pest management practices in a coordinated fashion.

Three things have received less attention during a collective action process for integrated pest management especially in the context of high value agri-food chains. Although a collective effort of integrated pest management approach allows certain crop losses; the effect is synergistic in the sense that it increases overall production and profits alongside reduced environmental pollution and human health risks. Secondly, localized adoption of integrated pest management among small group of farmers without considering the spatial dimension limits the success of the process since the pest can move to those non-adopted fields. Finally, pest management is not a static process controlling a single pest in a single season but it is a constant effort across temporal scales that require sustained collective action in a coordinated fashion.

Apart from the above, many other factors including socio-economic, institutional and cultural aspects and processes drive collective pest management action in farmers. Collective action (i) increase farmers' ability for joint investments and public-private partnership (ii) provide farmers with necessary information (iii) make possible vertical integration or contract farming [13, 21, 31-33].

Hence, future research should address the following issues:

- Characterization of public and private pesticides safety requirements of vegetables in national and international high value chains.
- Assessing the economic impact of pesticides safety standards of vegetables on smallholders in India.
- Estimating the impact of Indian vegetable producing smallholders' collective action on institutional and economic performance linked to pesticides safety.
- Comparing the pesticides safety performance among organic/GAP certified and noncertified vegetable growers of India.

3. MEASURING COLLECTIVE ACTION IN SMALL HOLDERS' VEGETABLE PRODUCTION

Depending on the empirical information, two approaches for performance analysis of smallholders' collectives are proposed here: (a) Ordinary Least Square (OLS) regression model [3] and (b) Data Envelopment analysis [34].

a. OLS Regression Model

$$Y = \alpha + \beta(group\ characteristics) + \gamma(institutional\ arrangements) + \delta(institutional\ \&\ economic\ environment) + \eta(ecological\ conditions) + \varepsilon, \varepsilon \sim N(0, \sigma^2) \quad (1)$$

Where Y is the dependent variable denoted by quantity of pesticides used by the individual farmers or farmers in the group, and other exploratory variables in the model are given in parenthesis.

b. Data Envelopment Analysis

Performance analysis or efficiency is a relative notion [35] and has its roots in production analysis. The efficiency concepts and its measurement can be applied to various fields including farm level production. Data Envelopment Analysis

(DEA) [34] is one such performance evaluation approach that identifies inefficiencies and also suggests possible improvements. This is also applicable to micro level decision management units (DMUs) [36] such as smallholders' collective or organisations [37]. Hence, the BCC-DEA[1] model under variable returns to scale which is proposed as [38]:

$$\min_{\Delta\lambda} \Delta j \ (2)$$

subject to

$$\sum_{j=1}^{n} y_j\lambda_j - y_j \geq 0; \lambda_j \geq 0 \ for \forall_j,$$

$$x_{ij}\Delta_j - \sum_{j=1}^{m} x_{ij}\lambda_j \geq 0,$$

$$\sum_{j=1}^{n} \lambda_j = 1, \ 0 \leq \Delta \leq 1$$

The assumptions of this model are as follows:

- Farm $j(j = 1,2, \ldots, n)$ produces a single output (y_j) using a combination of inputs

$$x_{ij}, where\ i = group\ charecteristics, institutional\ arrangements, ecological\ conditions, labor, machinery, seed, feretilisers\ etc.$$

- Input-oriented production frontier under variable returns to scale (VRS)

where (Δ_j) is a scalar which indicates the technical efficiency scores of the jth farm ($if\ \Delta_{j=}1$, then the farm is on the frontier and is technically efficient under variable returns to scale, but $if\ \Delta_{j=}\ 0$, then the farm lies below the frontier and is technical inefficient); (y_j) is a$(1 * n)$ vector of single output produced by the n farms; x_{ij}, x_{ij} is a $(m * n)$ input matrix; λ_j, λ is a $(n * 1)$ vector of weight value, $j\forall$ denotes for all j.

[1]Banker, Charnes and Cooper - Data Envelopment Analysis

From the Linear programing problem, assuming that the observed activities (x_j, y_j) (belong to production possibility set (P). Then minimize (Δ)that reduces the input vector (x) radically to (Δx)while remaining in P. The (Δ) is scalar representation of BCC- technical efficiency value, which varies in range of zero and one [36].

4. MEASURING ECONOMIC IMPACT OF PESTICIDES SAFETY

Along with production function analysis, damage control function approach is proposed for assessing the economic impact of production standards on farm revenue and pesticide usage [39]. 'Total revenue on sales of vegetables per acre per cropping season' is taken as the dependent variable, then the Cobb-Douglas production function with logistic damage control can be represented as:

$$ln(Q) = ln\,(\alpha) + \sum_{i=1}^{n} \beta_i ln(w_i) + \gamma G_i + ln[1 - \exp(\lambda - \alpha x_p)]^{-1} + v \quad (3)$$

where Q denotes total revenue per unit land received on sale of vegetables, the vector iW includes labor, fertilizer, seed, number of vegetable crops grown, other farm specific factors that affect total revenue such as, age of the household head and location-specific factors and iG denote adoption of Organic/IPM/GAP standards. The $i\beta's$ are the corresponding coefficients to be estimated; λ is constant and α is the parameter to be estimated for pesticides in logistic damage function framework.

CONCLUSION

The present paper has attempted to delineate briefly the pesticide safety issues in vegetable production in India. The collective action effort in food safety to increase environmental and economic performance of vegetable production by minimising pesticide risk is an emerging concept. In spite of the several advantages of collective action outlined above, methods of its measurement for ensuring food safety or pesticide risk reduction in high value chains in vegetable production of India has been hardly discussed. Hence, the methods presented above could be considered as a pioneering effort in this regard. Nevertheless, the absence of empirical evidence from India to test the presented models delimits importance of this paper. Yet, the paper could be a way forward for future empirical studies in this direction since these may augment the decision-making of all involved stakeholders minimizing welfare loss among smallholder vegetable production because currently little is known about most of the above issues which largely impede inclusion of India's smallholder vegetable farmers in global high value chains.

ACKNOWLEDGEMENTS

Declared None.

CONFLICT OF INTEREST

The author(s) confirm that this chapter contents have no conflict of interest.

REFERENCES

[1] Bignebat C, Codron J-M, Rouvière R. Monitoring safety in the fresh produce industry. INRA, France 2007.

[2] Bignebat C, Codron J-M. Organizational innovations and control of sanitary quality in the fruit and vegetable chain. INRA Sciences sociales. N°5-6 november 2006.

[3] Naziri D, Aubert M, Codron J-M, Loc NTT, Moustier P. Estimating the Impact of Small-Scale Farmers' Collective Action on Food Safety: The Case of Vegetables in Vietnam.Working Paper. Paris, France: CIRAD-MOISA 2010.

[4] Keikotlhaile BM, Spanoghe P, Steurbaut W. Effects of food processing on pesticide residues in fruits and vegetables: a meta-analysis approach. Food ChemToxicol 2010; 48(1):1-6.

[5] WHO (World Health Organisation). Public health impact of pesticides used in agriculture.WHO, Geneva 1990.

[6] Pimentel D, Culliney TW, Bashore T. Public health risks associated with pesticides and natural toxins in foods. In: Radcliffe EB, Hutchison WD, Eds. The Radcliffes's IPM World Textbook. University of Minnesota, St. Paul (Ed.) 1996.

[7] Dawson AH, Eddleston M, Senarathna L, Mohamed F, Gawarammana I. Acute Human Lethal Toxicity of Agricultural Pesticides: A Prospective Cohort Study. PLoS Med 2010; 7(10): e1000357. doi:10.1371/journal.pmed.1000357.

[8] Barzman M. Foresight prompts researchers in pest management to look beyond research. The Futures of Agriculture.Brief No. 05 - English. Rome: Global Forum on Agricultural Research (GFAR) 2012.

[9] Caswell JA, Johnson GV. Firm strategic response to food safety and nutrition regulation. In: J.A. Caswell, Ed. Economics of food safety. New York: Elsevier Science Publishing Company 1991.

[10] Henson S, Caswell JA Food safety regulation: An overview of contemporary issues. Food Policy 1999; 24(6):589-603.

[11] Martinez MG, Fearne A, Caswell JA, Henson S. Co-regulation as a possible model for food safety governance: Opportunities for public-private partnership. Food Policy 2007; 32(3): 299-314.

[12] Reardon T, Barrett CB, Berdegué JA, Swinnen J. Agrifood industry transformation and small farmers in developing countries. World Development 2009; 37(11): 1717-1727.

[13] Narrod C, Roy D, Okello J, Avendaño B, Rich K, Thorat A. Public–private partnerships and collective action in high value fruit and vegetable supply chains. Food Policy 2009; 34: 8-15.

[14] Ghosh SP. Deciduous Fruit Production in India.Deciduous fruit production in Asia and the Pacific 38, 1999. Available: ftp://ftp.fao.org/docrep/fao/004/AB985e/ab985e01.pdf, accessed 5th October 2012.

[15] Cao-Van P, Chau NM. Deciduous fruit production in Vietnam. Deciduous fruit production in Asia and the Pacific, 90, 1999. Available: http://www.applejournal.com/viet001.htm, accessed 5th October 2012.

[16] Ramanan G. Cost of production and capital productivity of grape Cultivation in Tamilnadu, India. Indian Streams Research Journal 2012; 2(1): 1-4.

[17] NHB (National Horticulture Board). Indian horticulture database-2014. NHB, Gurgaon 2014.

[18] Rajendran S. Environment And Health Aspects Of Pesticides Use In Indian Agriculture. In: Martin J Bunch, V Madha Suresh, T VasanthaKumaran, Eds. Proceedings of the Third International Conference on Environment and Health, Chennai, India 2003; pp. 353 – 373.

[19] Betne R. Indian Vegetables: Nutrition Pack with Toxic Cocktail. Toxics Link. Jan 2011. available: http://www.toxicslink.org/art-view.php?id=143, accessed 5th October 2012.

[20] Tixier P, De Bon H. Urban horticulture. In: R van Veenhuizen Ed. Cities farming for the future: Urban agriculture for green and productive cities. Ottawa: ETC-UA/RUAF, IDRC 2006.

[21] Roy D, Thorat A. Success in high value horticultural export markets for the small farmers: The case of Mahagrapes in India. World Development 2008. 36(10): 1874–1890.

[22] World Bank. India's Emergent Horticultural Exports: Addressing Sanitary and Phyto-Sanitary Standards and Other Challenges. South Asia Agriculture and Rural Development Department, World Bank, Washington D.C. 2006.

[23] Umali-Deininger D, Su M. Plenary paper prepared for presentation at the International Association of Agricultural Economists Conference, Gold Coast, Australia, August 12-18, 2006.

[24] Kumar A, Singh H, Kumar S, Mittal S. (2011).Value Chains of Agricultural Commodities and their Role in Food Security and Poverty Alleviation – A Synthesis. Agricultural Economics Research Review, 2011; 24: 169-181.

[25] World Bank. Food safety and agricultural health standards: Challenges and opportunities for developing countries exports. World Bank Sector Report No 31207, World Bank, Washington, DC 2005.

[26] Henson S, Reardon T. Private agri-food standards: Implications for food policy and the agri-food system. Food Policy 2005. 30: 241–253.

[27] Lee J, Gereffia G, Beauvais J. Global value chains and agri-food standards: Challenges and possibilities for smallholders in developing countries. PNAS, Early edition December 13, 2010. Available: www.pnas.org/cgi/doi/10.1073/pnas.0913714108, accessed 25th September, 2012.

[28] Bosc PM, Eychenne R, Hussein K, Losch B, Mercoiret MR, Rondot P, Macintosh-Walker S. The role of rural producer organizations in the World Bank rural development strategy, Rural Strategy Background Paper No 8, Washington, DC, World Bank, 2002.

[29] Winfree J, McCluskey JJ. Collective reputation and quality. American Journal of Agricultural Economics2005; 87(1): 206-213.

[30] Markelova H, Meinzen-Dick R, Hellin J, Dohrn S. Collective action for smallholder market access. Food Policy 2009; 34(1): 1-7.

[31] Berdegué JA, Balsevich F, Flores L, Reardon T. Central American supermarkets' private standards of quality and safety in procurement of fresh fruits and vegetables. Food Policy 2005; 30: 254-269.

[32] Henson S, Masakure O Boselie D. Private food safety and quality standards for fresh produce exporters: The case of HorticoAgrisystems, Zimbabwe. Food Policy 2005; 30(4): 371-384.

[33] Moustier P, Tam PTG, Anh DT, Binh VT, Loc NTT. The role of farmer organizations in supplying supermarkets with quality food in Vietnam. Food Policy 2010; 35(1): 69-78.

[34] Charnes A, Cooper WW, Rhodes E. Measuring the efficiency of decision making units. European Journal of Operational Research 1978; 2: 429-444.

[35] Coelli T, Rao DSP, Battese GE. An Introduction to Efficiency and Productivity Analysis. Kluwer Academic Publisher, London 1998.

[36] Cooper WW, Seiford LM, Tone K. Introduction to Data Envelopment Analysis and its uses. Springer Science and Business, New York 2006.

[37] Constant LA. Empirical Analysis of Agricultural Productivity: Growth in Benin and mainly factors which Influence growth. Australian Agricultural Economics and Resources Economics Society 55th Annual Conference-Melbourne, Australia 8-11 February 2011.

[38] Banker RD, Charnes A, Cooper WW. Some Models for Estimating Technical and Scale Inefficiencies in Data Envelopment Analysis. Management Science 1984); 30(9): 1078-1092.

[39] Lichtenberg E, Zilberman D. The econometrics of damage control: why specification matters. Amer J Agr Econ. 1986; 68:261-273.

Chapter 13: Documentation and Farmer Perception on Biodiversity in Rice Fields of Palakkad District

CHAPTER 13

Documentation and Farmer Perception on Biodiversity in Rice Fields of Palakkad District

Smitha K.P.[1,*] and Anilkumar A.[2]

[1]*Community Agrobiodiversity Centre, M.S.Swaminathan Research Foundation, Puthurvayal P.O, Kalpetta, Wayanad and* [2]*Dept of Agrl.Extension, College of Agriculture, Vellayani, Thiruvananthapuram*

Abstract: Rice wetland being highly fragile, the adverse effects of input intensive farming are prominent and far-reaching which questions the very existence of human race in this planet. Moreover, these practices are leading to extinction of many valuable flora and fauna. With the intension to study the effect of input intensive farming on the faunal and floral population associated with the rice fields, perception of farmers on enrichment of biodiversity by way of adopting eco-friendly farming was recorded. The investigation was undertaken in the Palakkad district, Kerala. A total of 40 beneficiary farmers were selected from the study area. The dimension was measured by using an interview schedule developed by the researcher for the purpose. To support this study on perception of farmers', major organisms constituting the rice ecosystem were identified and documented by taking photographs. Elderly farmers were contacted with these photos to collect information on the variation in the diversity over the years. The study reveals that majority of the respondents (62.5%) perceived biodiversity enrichment as a major social benefit of eco-friendly farming. Most of these farmers opined that reduced use of chemical inputs and avoiding poaching in eco-friendly farming leads to conservation of biodiversity in this eco-system. 37.5 per cent respondents attributed biodiversity changes to other factors like climate change and other human interventions, besides eco-friendly cultivation. A list of the flora and fauna seen in rice fields and its population variation over years was recorded as a part of the study. As high external input agriculture has been proven to be unsustainable, ensuring food security requires a shift towards low external input/eco-friendly agriculture, which would maintain input-output cycles as well as the ecosystem balance.

Keywords: Rice eco-system, biodiversity, Farmer perception, eco-friendly cultivation, flora and fauna.

1. INTRODUCTION

Rice can be grown under diverse ecological conditions and has a very prominent role in maintaining the microclimate of the area and provide shelter to diverse

***Corresponding author Smitha K.P.:** Senior Scientist, Community Agrobiodiversity Centre, M.S.Swaminathan Research Foundation, Puthurvayal P.O, Kalpetta, Wayanad; Tel: 09895090006; E-mail: smithavasanth9@gmail.com

Laladhas K.P., Oommen V. Oommen and Sudhakaran P.R. (Eds.)

flora and fauna. They have an imminent role in conserving water, stabilizing ground water table and preventing floods. The indiscriminate use of chemical fertilizers and pesticides has polluted the wetlands in the state extensively and has turned them into wastelands. Over 7,000 ha of paddy fields in the Kuttanad region, another "rice bowl" of Kerala, next to Palakkad is now lying fallow [1]. The adverse effects of conventional rice farming included depletion of ground water resources, fertilizer and pesticide contamination of soil and water, loss of biodiversity and genetic erosion. Maximum population of natural enemies was observed in natural biological control plot (NBCP) as compared to need based application of pesticides and scheduled based application of insecticides plot which itself shows the affect of chemical inputs on the faunal diversity of rice eco-system [2]. High mortality of spiders, ants, beetles and *Telenomus* sp. has been reported to be caused by pesticides [3, 4]. The chemical inputs have also affected the micro flora and fauna of this wetland ecosystem. Studies revealed the effects of the herbicide butachlor on the population of the *P. fluorescens* [5]. Increasing concentrations of all the herbicides from 0.5x to 2x was found to severely reduce the colony forming unit (c.f.u) counts of this bio agent.

As human beings, we have a moral responsibility for conserving the environment for holistic and inclusive growth with a long-term vision and perspective rather than satisfying short sighted and sectarian requirements in the pretext of development. The sustainability of the development will be at stake, if we follow a development paradigm neglecting the production of food grains and other basic livelihood requirements. It's high time we develop farming techniques for sustainable agriculture, considering the environmental, production and socio-economic dimensions. As high external input agriculture has been proven to be unsustainable, ensuring food security requires a shift towards low external input/eco-friendly agriculture, which would maintain input-output cycles and ecosystem balance. It is estimated that the output/input ratio in conventional farming is 3.76, while in the case of ecological farming it is more than 4.95 (Sharma, 2001). In the case of rice, which is a highly fragile wetland eco-system, the impacts of high external input farming are prominent and far-reaching. Considering the threats of intensive farming in rice, there is an urgent need to study in detail how farmer perceive the deterioration of biodiversity and which are the factors they consider important in the disintegration process. Success of any sustainable agriculture development effort is dependent on the farmers' perception on these issues.

The objectives is to:

1. Study the perception of respondents regarding biodiversity enrichment by adopting eco-friendly rice cultivation.

2. Documentation of the floral and faunal biodiversity variation over the years.

Enrichment of biodiversity in the study was defined as the increase in the population of flora and fauna in rice fields by way of adopting eco-friendly practices in rice cultivation.

2. METHODOLOGY

The research was undertaken in the Palakkad district, Kerala as the district accounts for about $1/3^{rd}$ of the total area under rice cultivation of Kerala state. Four blocks, Kollengode, Kuzhalmannam, Chittur and Alathur were selected purposively based on extent of area under rice. Thekkinchira, Nelliancaud, Manchira and Kolapadam *padashekarams* were randomly selected from Kollengode, Kuzhalmannam, Nalleppilly and Erimayur Grama Panchayats respectively (Fig. **1**). Ten beneficiary farmers were randomly selected from each *padashekaram*. Thus a total of 40 beneficiary farmers were selected from the study area. The selection process is detailed in the tree chart (Fig **2**). A direct survey approach was adopted for recording the primary data from the respondents at the field level, based on ex-post facto design. The interview schedule was pre-tested with 10 per cent of the non-sample respondents at random.

The dimension was measured using the schedule developed by the researcher for the purpose (Annexure-I). The schedule had two items, of which the first item had six sub items depicting the biodiversity in rice fields. It was rated by the respondent on a three-point continuum *viz.*, 'increased', 'no change' and 'decreased' with scores 3, 2 and 1 respectively. The second item was rated on a five point continuum namely 'very much increased', 'increased', 'no change', 'decreased' and 'very much decreased' with scores 5, 4, 3, 2 and 1 respectively. The summation of scores reflects the perception of respondents regarding biodiversity enrichment by adopting eco-friendly rice cultivation. The scores ranged from 7 to 23. For a comparison of biodiversity variation in the post green revolution period in contrast to the past, major organisms constituting the rice eco-system were identified and documented by taking photographs during the pilot study and discussion with experts working in the respective area and farmers. The data on biodiversity variation were collected by means of qualitative research method, the focus group discussion. Ten elderly farmers from the locations under study were called upon for the discussion with the photos of the flora and fauna of rice fields to collect information on the variation in the diversity over the years. There was a moderator who was the investigator and an assistant moderator to report the discussion who was one among the progressive farmers of the area to report the discussion. After a detailed discussion consensus regarding the variation in the population of the flora and fauna of rice eco system of the location was arrived at. The discussion was properly reported and was finally considered for qualitative analysis.

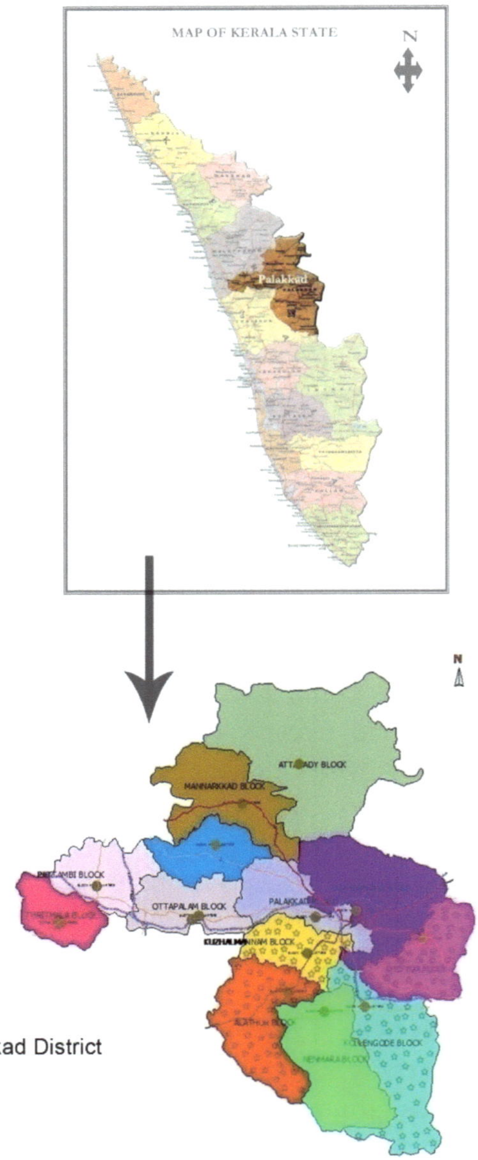

Palakkad District

Fig. 1 Maps Showing the locations of study

KOLLENGODE BLOCK

KUZHALMANNAM BLOCK

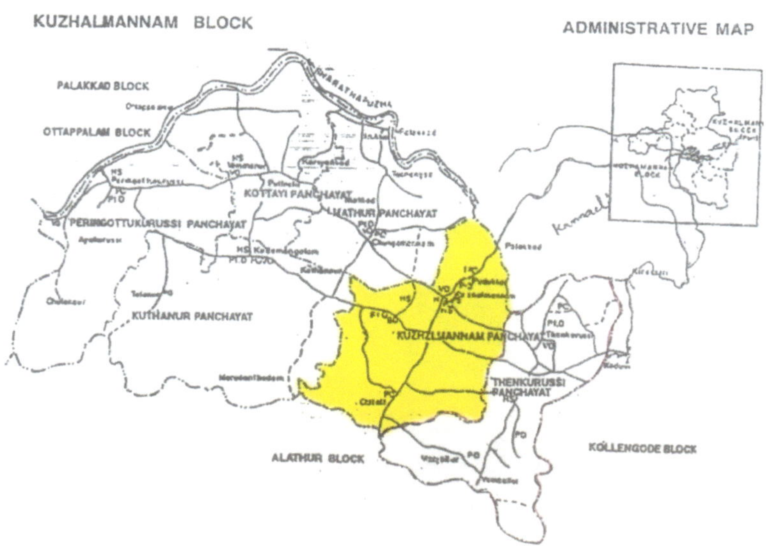

Maps Showing Kollengode & Kuzhalmannam Blocks

ALATHUR BLOCK

CHITTUR BLOCK

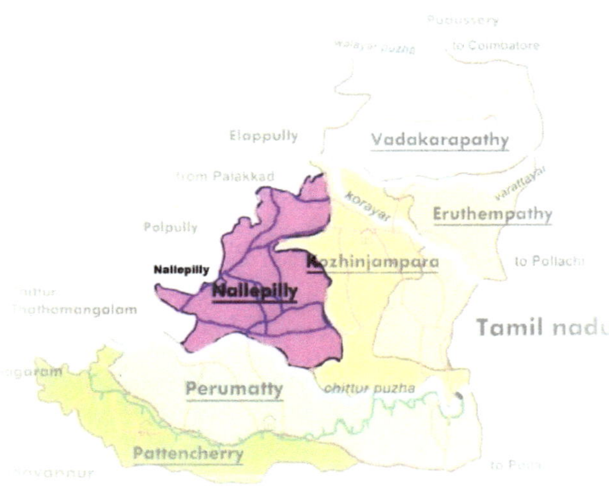

Maps Showing Alathur & Chittur Blocks

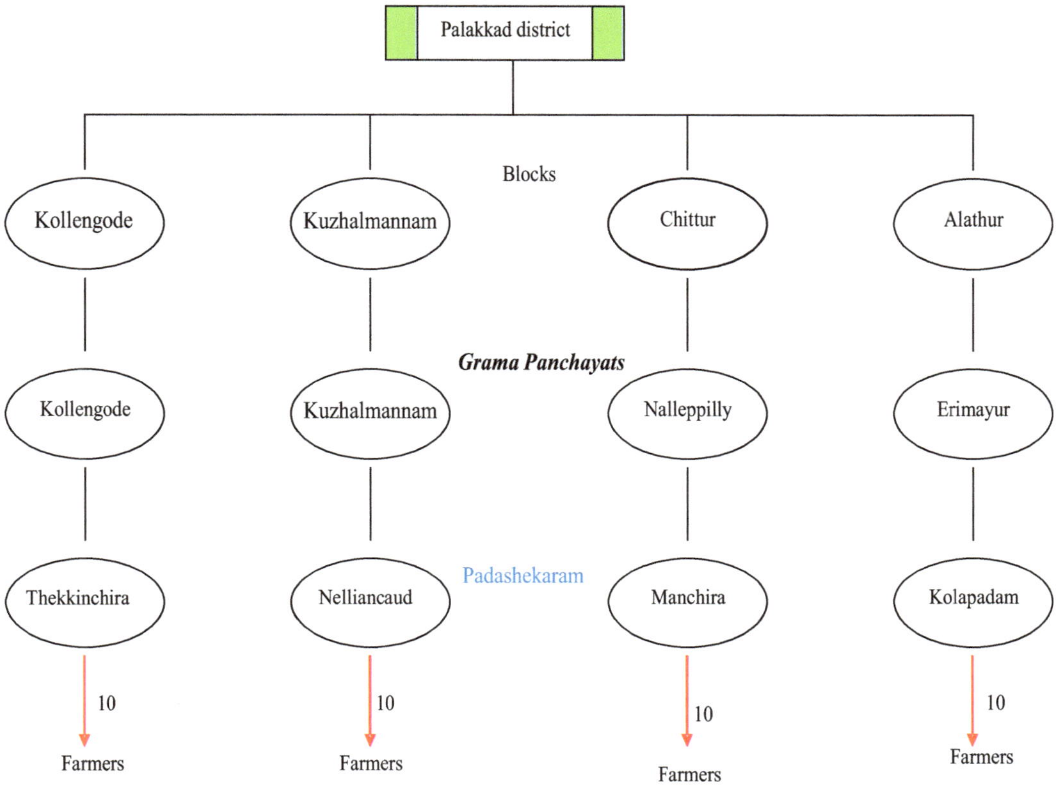

Figure 2: Selection of area/respondents.

RESULTS AND DISCUSSION

The results of the study are presented in Table **1**. Study reveals that majority of the farmer respondents (62.5%) perceived biodiversity enrichment as a major social benefit of eco-friendly farming. Most of the farmers opined that reduced use of chemical inputs and avoiding poaching in eco-friendly farming leads to conservation of biodiversity in this eco-system. 37.5 per cent respondents attributed biodiversity changes to other factors like climate change, development interventions like construction of buildings and roads, paddy land conversion, increased sand mining, poaching *etc*, rather than eco-friendly cultivation. Most of the farmers opined that natural enemy population, earthworms and other flora and fauna has increased in the fields managed in eco-friendly way. Maximum population of natural enemies has been found in natural biological control plot (NBCP) [2] and, synthetic pyrethroids produced more than 90 per cent mortality of ants, spiders, beetles and Telenomus sp in rice fields [4].

Table 1: Distribution of respondents into low and high perception categories on the social benefit of eco-friendly rice cultivation n = 40

Variable	Social benefit			
	Low (< mean)		**High (>or = mean)**	
	F	**%**	**F**	**%**
Enrichment of biodiversity	15	37.5	25	62.5

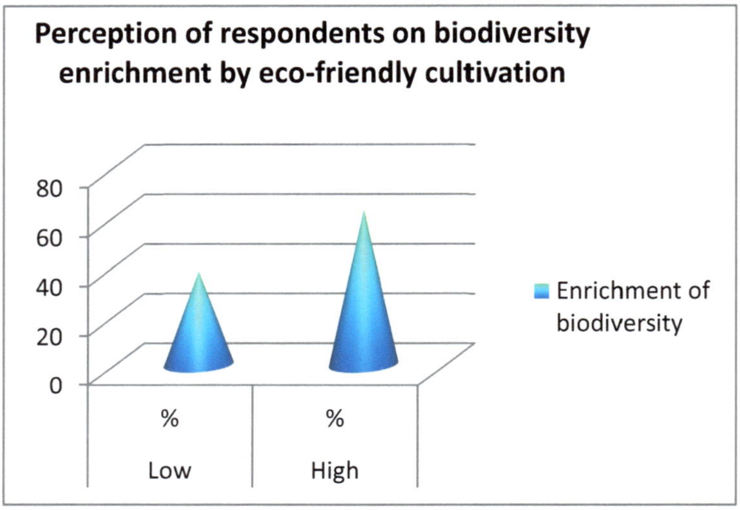

Graph-biodiversity enrichment

There has been a significant change in the faunal and floral diversity in rice fields over the years that may be attributed to very many reasons. Some of the reasons for reduction in diversity according to farmers say are the overuse of inorganic inputs and poaching. The results of the study are presented in Table **2**. Post green revolution period, inorganic fertilizers were indiscriminately used in the rice field by the farmers. Application of pesticides like endrin and folidol was extensive in late 70s, in the study area, which had reduced the population of frog, snakes, fish and eagle. Due to eutrophication caused by over use of fertilizers and contamination of water bodies there was a drastic reduction in the fish population in rice fields. Population of small weed fish (*paral*) and local *muzhi* has substantially reduced while that of African *muzhi*, which was introduced, has increased over the years.

A gradual reduction in natural enemy population was observed over the years in the study areas. Population of spiders, damselflies, dragonflies, ladybird beetles, grasshoppers, crickets, butterflies, rove beetles, water striders *etc.* has reduced. A drastic reduction in the population of frogs, water snakes, crabs, green lizards/chameleons *etc.* were reported by the farmers in the discussion. Earlier in the hillsides, near rice fields, turtles were seen in plenty whose population has drastically declined recently.

Table 2: Farmers' perception on variation in the floral and faunal diversity over years

Organisms whose population/menace has increased over the years				
Sl No.	**Flora**		**Fauna**	
	Common Name	**Scientific Name**	**Common Name**	**Scientific Name**
1.	Barnyard grass (*Mal. Kavada*)	*Echinochloa crusgalli*	Purple Moorhen (*Mal. Mayilkozhi*)	*Porphyrio porphyrio*
2.	Cyprus (*Mal. Manjakora)*	*Cyprus iria*	Cattle egrets *(Mal. Vella kokku)*	*Bubulcus ibis*
3.	Duck weed	*Lemna sp.*	Peacock (*Mal. Mayil*)	*Pavo cristatus*
4.	Hygrophila (*Mal.Vayalchulli*)	*Hygrophila auriculata*	Wild boar (*Mal. Kattu panni*)	Sus scrofa
5.	Water hyacinth (*Mal. Kulavazha*)	*Eichornia crassipes*	Deer (*Mal. Man*)	Cervus sp.
6.	*(Mal. Kora pullu)*	*Fimbristylis miliaceae*	Leech (*Mal. Atta*)	Haemadipsa sylvestris
6.	Chara *(Mal. Chandi)*	*Chara spp.*	Fish: African muzhi	*Clarias gariepinus*
7.	Spirogyra *(Mal. Payal)*	*Spirogyra spp.*		
Organisms whose population has decreased over the years				
1.	Hydrilla	*Hydrilla verticilata*	Spiders	Arachnida
2	Penny wort *(Mal. Brahmi)*	*Bacopa monnieri*	Damsel fly	Odonata: Zygoptera
3	Nymphaea (*Mal. Ambal)*)	*Nymphaea sp.*	Dragon fly	Odonata: Anisoptera
4	Wedelia *(Mal. Manjakkayonni)*	*Wedelia trilobata*	Lady bird beetle	Coleoptera: Coccinellidae
5	Glenwoodgrass *(Mal. Pollakala)*	*Saccolepis interrupta*	Grass hoppers	Acridoidea: Tettigonidae
6	Bloodgrass(*Mal.Changali/Naringa*)	*Iscsne miliaceae*	Crickets	Grylloidea
7	Puncture vine *(Mal. Njerinhil)*	*Tribulus terrestris*	Butterfly	Lepidoptera: Hesperidae
8			Rove beetles	Coleoptera: Staphylinidae

Table 2: contd…

9			Water striders	Hemiptera: Gerridae
10			Frogs	Anura
11			Water snakes	
12			Small weed fish *(Paral)*	*Puntius ticto, Puntius amphibius*
13			Local Muzhi/nadan muzhi	*Clarias dussumierii*
14			Crabs	
15			Garden lizard/chamaleon	
16			Grey herons	*Ardea cinerea*
17			Mynah	*Acridotheres tristis*

Avi-faunal population too has declined. Mynah, brown egrets *etc.* were some among the birds that has decreased in its number. Cattle egret has increased in number from that of the past. The problems of peacocks, deer and wild boar have increased in the Kollengode region in the hillsides of Sitarkundu. This might be due to the increased interference of human beings in their habitats in the hills and reduction in the predator population. There were even cases of finding the habitats of wild boar in the residential areas near rice fields. The problems of leech in these areas had also increased which in farmers' opinion were transmitted from hills through dam water which is now days used for irrigation purpose. Earlier leeches were not seen in rice fields, since the fields were then irrigated with water from traditional sources.

There had also been a shift in the floral diversity. Population of plants like hydrilla, brahmi, nymphaea, wedelia, glenwood grass, bloodgrass, puncture vine *etc.* has declined while that of eichornia, hygrophila, barnyardgrass, cyprus, lemna *etc.* had increased from that of the past as reported by the rice farmers of the study area. Interestingly in the discussion reduction in the population of puncture vine was associated to the reduction in livestock population especially goats in the surrounding areas as the nut lets disperse by adhering to the fur of grazing animals. Reduction in the population of Brahmi according to the farmer respondents was mainly due to conversion of wetlands and reduction of marshy areas which is the main habitat of the plant. Similar observation was given in the study on land use change [6]. All the farmers supported the opinion that medicinally important weedy species has reduced in numbers over the years whereas introduced plants like *Eichornia* and *Pistia* has become a menace in wetlands and waterbodies in Palakkad.

CONCLUSION

In this era of sustainable agriculture people have started realizing that, many of the environmental problems caused by farming are a direct result of an increasingly external input intensive and specialized agriculture and that any harm they cause to the rice eco-system entails irreversible transformation of the ecosystem. The principle farmers have to follow, to make sustainability a reality is "live and let live". Eco-friendly farming systems and techniques have been developed on the basis of a holistic view of human beings' existence within the biosphere and the awareness of their dependence on the natural resources. Thus, our goal should be to achieve an "ever-green revolution" in rice, which could improve productivity in perpetuity without associated ecological harm so that there is not only adequate rice in the market but also happiness on the faces of both farmers and consumers [7].

ACKNOWLEDGEMENTS

This paper has been possible with the sincere support of the rice farmers of Palakkad district.

The financial support from Kerala Agricultural University during the first author's doctoral research, of which this paper forms a part, is gratefully acknowledged.

CONFLICT OF INTEREST

The authors confirm that this chapter contents have no conflict of interest.

REFERENCES

[1] Nair PKP. Bailout rice in Kerala [online]. The Hindu Business Line Oct.10, 2008. Available:http://www.thehindubusinessline.com/2008/10/10/stories/2008101050340900.htm. [15 Dec 2009]

[2] Kumar S, Khan MA, Kumar A. Biodiversity and population trend of predatory fauna associated with major insect pests of rice. Green Farming J. 2008; 1(7): 36-37.

[3] Oi B. Population dynamics of spiders in rice fields and their effect in controlling rice plant hoppers. Tiangsu Agric.Sci. 1990; 4: 29-32.

[4] Chen ZF. Occurrence of Leptocorisa acuta (Thunberg) and its control. Entomological knowledge 1992; 29(6): 328-331.

[5] Bhanu C, Jain N, Bedi M. Effect of rice herbicides on bio control agent *P. fluorescens in vitro*. Proceedings of the second International Rice congress, 9-13 October, 2006, New Delhi 2006.

[6] Mathew JC. Land use change analysis of three selected panchayats in Kuttanad using Remote Sensing data and GIS - Report. Centre for Environment and Development, Thiruvananthapuram 1999.

[7] Sharma AK.A handbook of organic farming. Jodhpur, India: Agrobios, 2001. Swaminathan MS. Science and shaping the future. Proceedings of the second International Rice congress, 9-13 October 2006, New Delhi. pp. 6-13.

ANNEXURE - I

Questionnaire for farmer perception on enrichment of biodiversity in eco-friendly rice cultivation

a. What do you think has happened to the biodiversity status of rice fields where eco-friendly cultivation practices are adopted?

Biodiversity	Increased	No change	Decreased
1. Medicinal plants			
2. Other weed flora			
3. Natural enemies			
4. Aquatic organisms			
5. Birds			
6. Other animals			

b. What do you think is the extent of cultivation of traditional varieties in eco-friendly rice cultivation?

Very much increased/Increased/No change/Decreased/Very much decreased

Chapter 14: An Initiative towards Achieving the 'Access and Benefit Sharing' Provisions of Biodiversity Act and Rules 2004, India 2002

CHAPTER 14

An Initiative towards Achieving the 'Access and Benefit Sharing' Provisions of Biodiversity Act 2002 and Rules 2004, India

Sudipta Mukherjee, Soumyendra Nath Ghosh* and Debal Ray

West Bengal Biodiversity Board, Poura Bhawan, 4th Floor, FD-415A, Sector – III, Salt Lake City, Kolkata, India

Abstract: One of the major concepts of conservation of biodiversity on the International and National level is "equitable sharing of benefits arising out of uses of genetic resources". The main instrument for ensuring this in India is documenting biodiversity along with traditional knowledge in the form of People's Biodiversity Register (PBR). The PBR database reveals that a large number of Traditional Rice Varieties are being maintained, till date, by the farmers on their own. It is reported that 467 traditional varieties are still in cultivation in the fields of West Bengal, against a number of nearly 5000 that existed 50 years back in the state. Protection of Plant Varieties and Farmers' Right (PPV&FR) Act 2001 [1] offers rights to the farmers that is akin to some degree to Biological Diversity Act, 2002. West Bengal Biodiversity Board has taken initiative to build synergy between these two Acts. As per Section 14-19 of the PPV & FR Act, the process of registration of varieties in the name of concerned farmer/group of farmers from different parts of the state has been initiated which would provide them a right to get adequate compensation from those who develop the new varieties for commercial benefits using the traditional ones. These two complementary acts may change the scenario of "Access and Benefit Sharing" provisions, particularly in the agricultural sector.

Keywords: Access Benefit Sharing, Biodiversity Act, CBD, equitable sharing, farmer's varieties, People's Biodiversity Register, PPV& FR Act, Rice, traditional knowledge.

1. INTRODUCTION

The Convention on Biological Diversity, 1992 recognizes three basic principles of Biodiversity Conservation: a) Conservation of Bio-resources, b) its sustainable utilization and c) equitable sharing of the benefits. All the national legislations

***Corresponding author Soumyendra Nath Ghosh:** West Bengal Biodiversity Board, Poura Bhawan, 4th Floor, FD-415A, Sector – III, Salt Lake City, Kolkata, India; Tel: +91 33 2358 2731; Fax: +91 33 2358 2763; E-mail: sro.wbbb@nic.in, ms.wbbb@nic.in

Laladhas K.P., Oommen V. Oommen and Sudhakaran P.R. (Eds.)

that followed, recognized the above three principles of biodiversity conservation. While there are many more legislations dealing with the first two goals of CBD-like The Forest Act (1927), The Indian Fisheries Act, The Wildlife (Protection) Act, 1972 *etc.*–the third principle, *i.e.* equitable sharing of benefit has received relatively little attention till recently. However, this principle is a unique contribution of the CBD to the field of natural resource management.

According to The Biological Diversity Act, 2002, individuals, institutes or organizations intending to utilize bio-resource and/or access the traditional knowledge have to share benefits with the Biodiversity Management Committees/State Biodiversity Boards. The resultant funds generated at local or state level are then divided among the stakeholders for conservation activities. West Bengal Biodiversity Board, established in the year 2004, primarily deals with implementation of The Biological Diversity Act, 2002 and the Biological Diversity Rules, 2004. The Board is dedicated in enforcing equitable benefit sharing of the resources through a legal instrument called People's Biodiversity Register (PBR), where the local bio-resources and the traditional knowledge associated with those are being recorded.

While implementing the above, the West Bengal Biodiversity Board noticed that a synergy can be developed with The Biological Diversity Act, 2002 and The Protection of Plant Varieties & Farmers' Rights Act[2], 2001 [1]. As a signatory of General Agreement on Tariffs and Trade (GATT), the Govt. of India introduced PPV&FRA Act, first of its kind in the world to recognize and protect the rights of farmers in respect to their contributions in conserving traditional varieties of crops. Anyone willing to derive benefits out of any traditional plant variety registered in the name of a farmer has to share the benefit with the farmer before accessing it. Thus, there is a similarity between the Biological Diversity Act and PPV&FR Act when it comes to the implementation of the access and benefit sharing principles.

2. EQUITABLE SHARING OF BENEFITS

Preparation of PBR lays the basics for access and benefit sharing. The location and the associated traditional knowledge of bio resources documented in PBR gives prospective user/buyer an idea whether his needs may be met from the area. This leads to a legal agreement between buyer and the Biodiversity Management Committee (BMC) and transfer of the resource against a specified benefit.

The benefit-sharing principle for farmer's varieties is more or less the same, although details differ. The specific and detailed information required for farmer's variety registration was acquired through existing literature from different sources, from experts in the field, direct observations from the farmers' fields and from ongoing PBR exercise. Various characteristics of a rice variety at different growing stages were recorded according to the PPV&FRA specified guidelines [2].

Various characteristics of a rice variety at different growing stages were recorded according to the PPV&FRA specified guidelines [2].

Information regarding cultivation of traditional rice varieties along with the associated traditional knowledge have been collected from analyzing 42 PBRs (up to 2012), many of which are lacking in existing literature. Special emphases were given on areas experiencing stressed conditions, such as Bankura, Purulia, Birbhum, Jalpaiguri and Sundarbans. Many of the rice varieties cultivated there are stress tolerant. Thus, if the BMCs are empowered by imparting negotiation skill and some knowledge on legal instruments (MAT, MTA, PIC) they may be able to market their product at fair prices. The State Board and National Biodiversity Authority (NBA) are building capacity in this direction.

Panicles of some traditional rice varieties
1. Kalam, 2. Kartiksaal, 3. Lal Challisha, 4. Malsira, 5. Masuri Paijam, 6. Nepali Paijam

Figure 1: Panicles of some of the traditional rice varieties.

According to the data collected so far from the PBR, there are more than 100 traditional rice varieties being cultivated in the state of West Bengal spread over different districts. As on end of 2012, DUS characterization of 23 varieties of rice (Table **1**) & (Map **1**) were done and sent to the PPV&FR Authority for registrations.

Table 1: Rice varieties and their characteristics

Name of the Variety	Localities	District	Name of the Farmer's Group	Special Character
Bhutia	Different villages of Dantan – II Block	Paschim Medinipur	Jankapur Janokalyan Samobay Krishi Unnayan Samiti Limited	Brown rice
Dhulia				Good table rice
Gayabali				Aromatic rice
Nimisaal				
Mugi				
Modra				
Raygarh				Flood tolerant
Kokhuria				
Nadna				
Bhasamanik	In village Mankanali	Bankura	Khiraijuri MalindasiS. K. U. S. Ltd.	Flood tolerant
Danaguri				Aromatic rice
Bhutmuri				Drought tolerant
Kashiphul				Drought tolerant
Asanlaya				Drought tolerant
Tulaipanji	Different villages near Mohinigunj	Uttar Dinajpur	Basian Samabay Krishi Unnayan Samity Ltd.	Aromatic rice
Chamormoni	Villages of Joynagar – II and Mandirbazar Block	South 24 Parganas	Lokamata Rani Rashmoni Mission	Good table rice
Kanakchur				Aromatic rice
Dudheswar				Aromatic rice
Marichsaal				Good for making puffed rice

Having previous experience and encouragement this process is being continued by the Board. More than 30 other farmer's varieties of rice have been identified from different districts of the state and documents of required characteristics and information of those, as per DUS guidelines [2] is going on. The new varieties are from Bankura, Purba Medinipur, Jalpaiguri and South 24 Parganas districts of the state. Most varieties are being collected from areas that experience environmental stress which eventually leads to development of different tolerant capabilities. For example, some varieties can grow in saline water where most of the HYVs cannot grow, whereas there are some varieties that show non-lodging characteristics even under flooded conditions.

Map 1. Districts covered during the first phase of documentation and registration of traditional rice varieties of West Bengal.

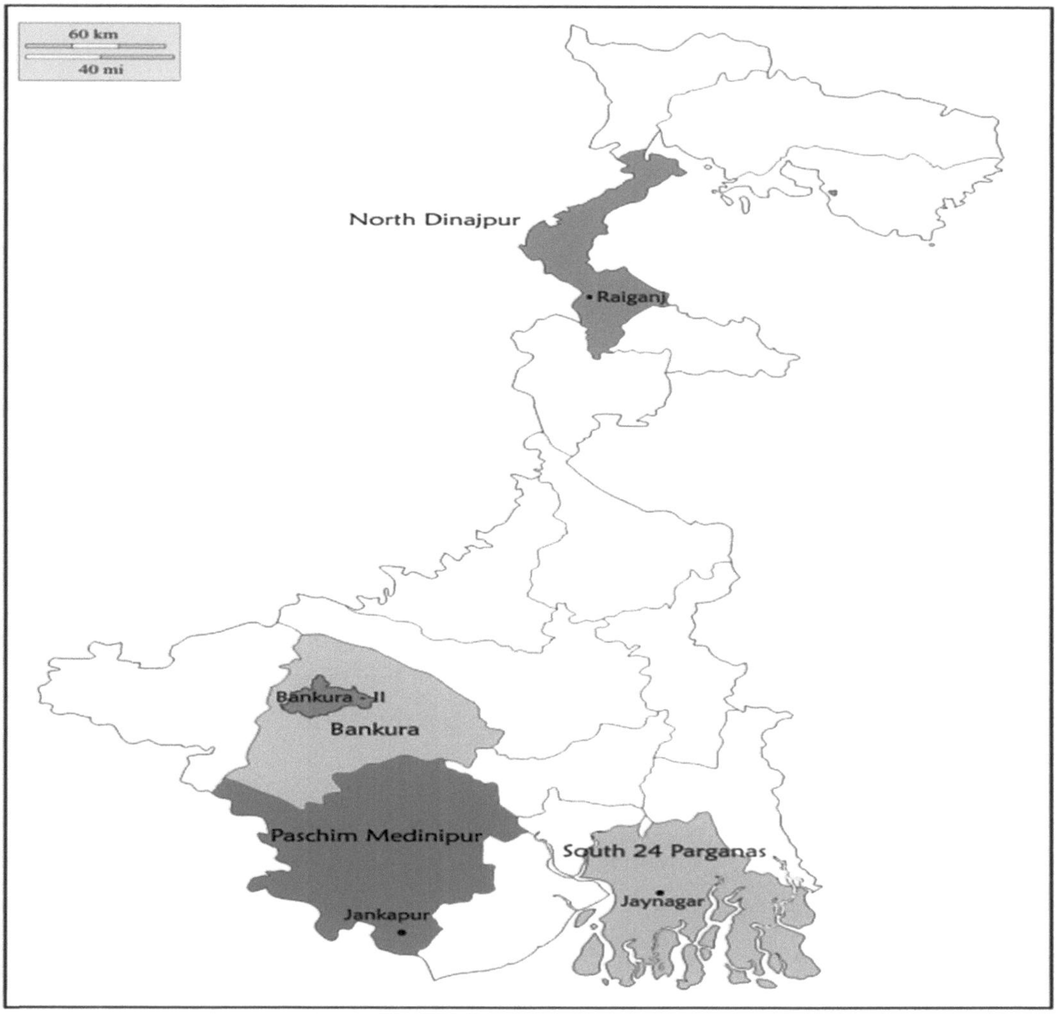

As per PPV&FR Act, getting the registration means, the farmer or the farmer group will get sole right to sell seeds in the market and also to bargain with the person or organization who wish to develop and market new superior varieties using these traditional varieties.

These two complementary acts The Biological Diversity Act, 2002 and The Protection of Plant Varieties and Farmers' Rights Act, 2001, if implemented properly as per their spirit, may change the scenario of "Access and Benefit Sharing" provisions, and will bring about new stimulus in agrodiversity conservation.

ACKNOWLEDGEMENTS

The authors want to convey sincere gratitude towards the Department of Environment, Government of West Bengal, for financing different initiatives to make documentation of traditional rice varieties in West Bengal possible. The authors also thank Prof. A. K. Sharma, Former Chairman and Dr. Asok Kanti Sanyal, Chairman, West Bengal Biodiversity Board for their encouragement. Lastly, the above mentioned initiatives would not have been successful without the active participation of people at grass root level.

CONFLICT OF INTEREST

The authors confirm that this chapter contents have no conflict of interest.

REFERENCES

[1] Ministry of Law, Justice And Company Affairs (Legislative Department).The Protection of Plant Varieties and Farmers' Rights Act, 2001. The Protection of Plant Varieties and Farmers' Rights Authority 2001.

[2] Protection of Plant Varieties and Farmers' Rights Authority (PPV & FRA). Guidelines for the Conduct of Test for Distinctiveness, Uniformity and Stability On *(Oryza sativa* L.). New Delhi: Registrar, on behalf of the Chairperson, PPV & FR Authority 2007.

CHAPTER 15

Actionable Innovative Financial and Economic Instruments for Biodiversity Conservation in India

Damodaran A.[*]

Indian Institute of Management, Bangalore, Karnataka, India

Abstract: One of the key elements of the Strategic Plan on Biodiversity Conservation and the Aichi Targets adopted by the Conference of Parties in the Tenth Session at Aichi Nagoya has been the issue of resource mobilization for Biodiversity Conservation. Resource mobilization for biodiversity conservation calls for deployment of economic instruments that mobilize resources and /or secure corrective behavior that facilitates biodiversity conservation in mega-diversity countries such as India. The paper looks at the concept of economic incentives and its role in biodiversity conservation through creative use of economic instruments. The backdrop chosen for the analysis is India. After delving into the different types of economic incentives in relation to their biodiversity impacts, the paper looks at how different fiscal and non fiscal instruments, including taxes, cess and related imposts can be combined with price and non price incentives and dedicated funds to create conditions for promoting conservation and sustainable utilization of biodiversity resources. The main argument advanced in this paper is that, apart from dedicated environmental fiscal and non fiscal instruments, it is possible to utilize existing fiscal instruments, including conventional direct and indirect taxes, by re-orienting their focus to securing positive biodiversity outcomes. The paper summarizes its findings by linking these concepts to valuation exercises conducted under the auspices of The Economics of Ecosystems and Biodiversity (TEEB) in India and elsewhere and the proposed novel trading instruments such as Biodiversity Off sets proposed in various economic and environmental forums.

Keywords: Biodiversity conservation, biological resources, cess, economic instruments, financing mechanisms, impost, India, indirect taxes, seigniorage, sops, tax.

1. INTRODUCTION

Financing mechanisms for biodiversity conservation include economic instruments that generate revenues, flow of resources as well as funds that lodge these resources. Both components afford considerable scope for innovative approaches. In the case of biodiversity conservation, innovative approaches are

***Corresponding author Damodaran A.:** Indian Institute of Management, Bangalore, Karnataka, India; Tel\Fax: +91-80-25278562; E-mail: damodaran@iimb.ernet.in

Laladhas K.P., Oommen V. Oommen and Sudhakaran P.R. (Eds.)

not only about creating 'new instruments' or 'new funds' with novel design features, but also about refashioning existing instruments to serve the cause of biodiversity conservation. After providing a background and delving into issues related to typologies of economic instruments, in the first three sections, the paper proceeds to consider existing ones like the water cess which has been set up as an environmental economic instrument in India. The paper then proceeds to look at direct and indirect taxes in India that can be re-tuned to achieve biodiversity conservation goals. The paper concludes by examining the National Biodiversity Fund and the National Green Mission 'Fund' that seek to address biodiversity relevant activities in India.

The important point about global environmental financing mechanisms in the context of global public goods like biodiversity is that it has to be based on 'new, additional and predictable' sources of financing. However in a country context, biodiversity is a 'national resource of international significance'. Apart from emphasizing on the criteria of 'additionality', the focus of financing biodiversity conservation in India has been on resolving externalities that lead to biodiversity loss. This can be done by introducing new economic instruments or re-designing existing fiscal instruments that do not have a 'biodiversity' focus at present.

The theory of externalities has been popularly employed by Welfare and Environmental Economists (aligned to the Welfare Economic School) to explain the loss of biodiversity. At a broad level, externalities are caused due to the fact that the benefits of biodiversity conservation are public and not private. Welfare economists argue that since the social value of an ecosystem exceeds its private or market value, an externality in the shape of 'non-accounted positive social values/negative values' exists, resulting in society not being able to penalize biodiversity endangering activities. Externalities, or the gap between private or social costs arise in ecosystems since complexities of life support systems and services are not precisely understood in scientific terms. This may arise due to non-perception of secondary values created by ecosystems and the energetics and patterns of inter-linkages between components of an ecosystem. This may also arise due to the difficulties in measuring functional diversity *i.e.* the range of species capable of performing particular functions under different environmental conditions. Alternatively, there could be lack of information of diversity of ecological functions performed by different species which requires measurement of number of species required for a given process [1].

The inception of TEEB in 2007, to structure and recommend appropriate valuation methodologies for different contexts, has succeeded in removing information gaps on ecological functions that partake of biodiversity related environmental services provided by ecosystems [2]. The ongoing TEEB exercise in India offers the potential to rework on certain existing instruments and introduce new instruments or mechanisms that can aid conservation of biological diversity in India.

Further, the reality is that the water scarce areas of rural India clearly demonstrate the importance of coupling adaptation to climate stress with bio-diversity conservation programmes [3].

2. BACKDROP

Global concern for conservation of the planet's biological diversity has its roots in the UN Conference on the Human Environment held in Stockholm in 1972. The "World Conservation Strategy" subsequently formulated marked the beginning of a series of steps which was to culminate in the Global Convention on Biological Diversity, that was opened for signatures at the 1992 United Nations Conference on Environment and Development held at Rio de Janeiro. This Convention entered into force on 29 December, 1993.

Convention of Biological Diversity is a milestone as it was for the first time that the global community stated its commitment towards conservation and sustainable use of bio-diversity, with a matching concern for sharing the benefits arising from utilisation of genetic resources with countries of origin. The three principal objectives of the Convention, captured in Article 1, reflect this spirit. The main objectives of the convention are:

- The conservation of biological diversity;
- The sustainable use of the components of bio-diversity;
- The fair and equitable sharing of benefits arising out of the utilization of genetic resources.

The objective of equitable sharing of benefits was fortified in the Convention, by parties reaffirming sovereign rights of states over their bio-diversity wealth, a principle which entitled countries with biological wealth to seek equity and fairness in sharing benefits arising out of the utilization of their resources. However the pre-condition for generating equitable sharing of benefits is conservation of biological resources by developing countries which harbour the cream of these resources. Conservation of biological resources can be achieved by strong legislations and enforcement of these legislations through policy and programme framework. The action for conservation has traditionally been presaged on the principles of command-and-control systems whereby biodiversity reserves are declared and the functioning of these reserves are guaranteed through strong steps for policing these reserves with the aid of punitive provisions of the legislations. In India biodiversity reserves have been constituted in the shape of forest lands, protected areas and other reserved public lands. The Indian Forest Act, the Wildlife Protection Act 1972, and the Environment Protection Act 1985 have been the principal legislations in India which have attempted to cover biodiversity and biodiversity reserves. However the first two legislations cover only a portion of the country's biodiversity wealth. The introduction of the National Biodiversity Act in 2002 succeeded in overcoming the deficiencies in the forests and environmental legislations that preceded this Act. The act combines a regulatory approach on access with a system of economic incentives for conservation of biodiversity through a holistic approach [4].

India's National Biodiversity Action Plan (NBAP) 2008 has sought to integrate biodiversity concerns into various sectors as one of its main process goals. Para 4.10 of the NBAP emphasizes the importance of valuation of goods and services provided by biodiversity and use of economic instruments in decision making processes [5].

India's decision to adopt economic incentives and instruments for bio-diversity conservation needs to be viewed against the backdrop of para 4.10 of the NBAP 2008. There is a need to carefully evaluate the likely impact of the various economic instruments for biodiversity conservation in the Indian context.

Given the background, this paper attempts to survey the following issues:

Existing economic instruments for environmental conservation and its utility to biodiversity conservation.

Feasibility of re-orienting certain economic instruments. These instruments may include forestry related charges and taxes, amongst others apart from non-dedicated instruments such as Income Tax, Corporate Income Tax, Property Tax, Agricultural Income Tax, Excise duties and Sales Tax, Value Added Tax

- Feasibility of introducing 'new' and 'innovative' economic mechanisms.
- The Biodiversity Fund and the Green India Mission 'Fund' [6].

3. ECONOMIC INCENTIVES AND INSTRUMENTS: CONCEPTS AND TYPOLOGY

Economic instruments act as vehicles for conferring economic incentives. The repertoire of economic instruments available for environmental purposes is impressive – so is the range of economic incentives that can be designed for biodiversity conservation. In a broad sense, economic instruments can be divided into two categories *viz.* 'imposts' and 'sops'. While 'imposts' can assume the forms of taxes, 'sops' can assume the forms of 'subsidy', 'allowances', 'funds for capital assistance', 'facilities for improved market' and 'return from benefits by way of use of *in situ* biodiversity' [7]. Apart from their range, economic instruments can be further categorized into two, *viz.* those which are 'dedicated' and specifically directed for conservation and those not 'dedicated' for conservation, but which enjoy the potential as a biodiversity conserving instrument, if refashioned. A fair amount of literature has been developed on incentives and economic instruments for environmental protection [8]. Though these primarily deal with economic instruments in developed country contexts, there are interesting examples from developing countries as well.

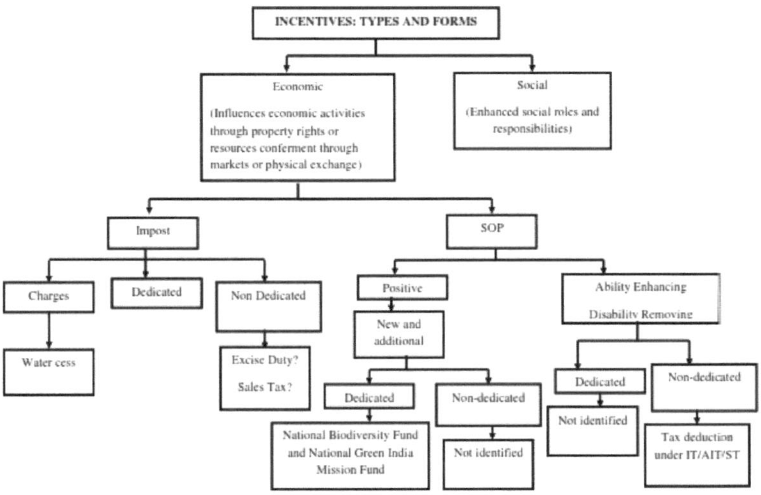

Figure 1: Incentives / instruments: types and forms.

4. PROFILE OF DIFFERENT ECONOMIC INSTRUMENTS IN INDIA

4.1. Water Cess/Charges

User charges are levied on the use of a specific goods or service and thereby on the accruing benefits that economic agents receives from such use [9]. Thus user charges allow costs of the goods or services to be distributed according to the benefit received. The benefit principle of public finance is implicit in charges. User charges are efficiently levied on publicly provided goods or services that enhance economic efficiency and have the characteristic that their users may be easily identified and excluded from their consumption. User charges should reflect marginal cost of service in case it is service oriented.

A 'cess' is a charge generally employed as a levy for some special administrative expenses which the name (health cess, education cess, road cess *etc.*) would indicate. When levied as an increment to any existing tax, the name matters not, because the validity must be judged in the same way as the validity of the tax to which it is an increment. Similarly, the word 'duty' is defined as "a payment to be public revenue levied upon the import, export, manufacture, or sale of certain commodities". 'Tax', in legal parlance, also means nothing more than money which individuals are compelled to pay for public purposes.

User charges have been pervasively applied in many countries including India. India's Water Cess Act is an outstanding example of an economic instrument employed on use of water as a natural resource. The Act provides for the levy and collection of a cess on water consumed by persons carrying on certain industries and by local authorities, with a view to augment the resources of the Central Board and the State Boards for the prevention and control of water pollution constituted under the Water (Prevention and Control of Pollution) Act, 1974. The target groups of the Water Cess Act are industry and local authorities. Such of those persons carrying on specified industry, as given in Schedule I of the Act are liable to pay the cess. The cess payable is linked with the water consumed by the industrialists or the local authority for the purposes specified in Schedule II of the Act. The rate is again to be specified by the Central Government. A ceiling for such rate is prescribed under the Act itself by the entry in column (2) of Schedule II. The machinery for the quantification of the cess payable is furnished by Sections 4 to 6. An obligation to furnish the return is cast on the persons liable to pay the cess. Where any person or local authority, liable to pay the cess under this Act, installs any plant for the treatment of sewage or trade effluent, such person or local authority shall, from such date as may be prescribed, be entitled to a rebate of {twenty percent} of the cess payable by such person or, as the case may be, local authority:

The proceeds of the cess is first credited to the Consolidated Fund of India and the Central Government may, if Parliament, by appropriation made by law in this behalf, so provides, pay to the Central Board and every State Board, from time to time, from out of such proceeds, after deducting the expenses on collection, such sums of money as it may think fit. While determining the sum of money to be paid to any State Board under this section, the Central Government considers the amount of cess collected by the State Government concerned under sub-section (4) of Section 6. There is a penalty for non-payment of cess within the specified time.

The water cess can be noted as a hybrid instrument which seeks to achieve control over use of water resources as well as achieve regulation of water pollution through its 'rebate provisions'. The significance of this instrument to biodiversity conservation is immense. The riverine ecosystems of India, are under threat of losing their biodiversity potential due to high levels of industrial and municipal pollution. India's wetlands, mangroves and coral reefs are also threatened by the increasing loads of organic and in-organic pollution. As India's National Biodiversity Action Plan (1999/2008) [5] notes, mangroves in and around Mumbai and Cochin have been subjected to the impact of industrial and sewage effluents. To this extent, the water cess is a valuable instrument in conservation of biodiversity. However the existing structure of water cess is not 'receiving stream based'. It is 'source based'. To this extent the water cess cannot be considered to have a biodiversity conservation bias. If industrial and non-industrial units that discharge their effluents into wetlands, coral reefs and mangroves are subjected to a special surcharge over the regular water cess rates, it will be in the interests of biodiversity conservation. This will refashion water cess as an instrument of biodiversity conservation as well. Alternatively a Pigouvian tax may be thought of for select 'air' and 'watersheds' which are rich in biodiversity. In short, the water cess has the potential to be refashioned as an instrument for biodiversity conservation. It can be 'habitat' specific. The introduction of mass based effluent/emission standards can save India's wetlands from the grip of local authorities and industries which source water from these significant ecosystems.

4.2. Other Charges for Biodiversity Conservation

4.2.1. Forest Seigniorage (Fee) in India

In the forests lands of India, permits are granted for removal of trees or other forest produce on seigniorage fees. Such grant is given if it is applied for bonafide agricultural or domestic use and other works of public utilities. Seigniorage fees apply to local demands for forest produce by local inhabitants enjoying certain privileges in forests, provided these demands are for bonafide domestic use. The seigniorage fees are regulated by State Forest Acts.

The seigniorage rate of tree wood or timber (with bark) and other forest produce for the State of Karnataka from any District or Reserved Forest are given in the Table **1**. The royalty levied on minor minerals in forest areas are also given in the table.

Table 1: Seigniorage rate of tree wood or timber (with bark) and other forest produce for the State of Karnataka

Forest Produce	When Disposed of to
Timber, Fire wood, grass, charcoal, Eucalyptus	Industries
Bamboo, reeds and canes	Pulp and paper industries except Cottage industries
Sandal wood	Sandal wood oil factories and others except artisans, religious institutions and cottage industries
Minor Forest produce not covered in the above two categories	Industries except large scale Multipurpose co-op societies

In Assam, royalty and taxes on forest produce are fixed by the government after scrutiny of market rates of timber in the state as well as rates in the neighbouring states. These rates are periodically revised. Sand stone and other minor forest produces are disposed of by Mohal system. The mohals are constituted for one or two year's duration wherein the quantity extracted and the declared area of mohal is specified in the contract. The mohals are sold on a competitive sales basis by a tender system and kists (instalments) are fixed in respect of tendered amounts. The mohaldar is responsible for harvesting and detection of illegalities and payment of kist money to Government as per agreement.

For sustainable management of forest resources, it is essential that a system exists for disposal of forest produce on the basis of prepaid licenses. The disposal is worked according to the prescriptions of sanctioned working plans. But when the resources were plentiful, the issue of prepaid licenses for firewood and other bona-fide use of forest produce on payment of seigniorage value were acceptable. But with more demand for these resources, this practice has created havoc.

4.2.2. Entry Fee in Relation to Wildlife Parks and Sanctuaries

Entry fee to wildlife parks, state zoos, elephant rides, and photography fees are fixed and collected by State Government or National Park Authorities as per scheduled rates. The efficiency of the instruments is of a high order as collections are high. However the effectiveness of the entry fee in encouraging sustainable tourism is questionable. Willingness to pay on the part of tourists to wildlife parks are higher than the modest entry fee prescribed and enforced.

4.2.3. Timber Auctions and Tendering Systems of Forest Departments

In States like Assam in India, timber extracted from Reserved Forests is sold in competitive market by tender system. Timber price varies according to species, size, quality market and accessibility. Beside tendered value, bidders have to pay Assam Finance Tax at 8% and Income Tax at 5.43%.

Generally no difficulties are observed in collection of these taxes and charges as the produce is released only after collection of amount. The only problem arises when the bidders quote high prices and do not pay the price they bid for to lift the materials. The forest produce has again to be put up for sale and the difference in royalty recovered from the defaulter. This delay leads to deterioration of the produce and waste of scarce timber resources. On the contrary, by insisting on a high net worth on the part of the bidders, it will be possible to ensure that bidders are genuine and wastage of timber does not result. Indeed a high rate of Assam Finance Tax and Income Tax may be levied for rare and quality timber and for high level bid prices (say, Indian Rupees 1000/- and above) so that quality timber cease to attract auction bidders. To this extent, there exists potential to discourage harvesting of 'quality timber', with positive consequences for biodiversity.

4.2.4. Forest Royalties

Forest royalties are levied on forest resources leased out to industries or other users. For example, bamboo areas are leased out to paper mills which in turn harvest bamboos within the leased area in return for royalties paid. In levying royalties the implicit market value of the leased resources is very often not taken into account. This results in accelerated depletion of forest resources. Further a volume or quantity based charge can leave un-harvested species that have lower market values. Enhanced royalties on forest resources can reverse depletion of biodiversity, thus contributing to sustainable consumption [10]. It needs to be noted that enhancement of forest royalties do not amount to levy of tax, since forest royalties are not public purpose levies and therefore are beneficial only to the forest sector. Hence, it has the potential to achieve the purpose of a dedicated economic instrument for biodiversity protection.

4.2.5. Taxes

Environmental taxes are dedicated tax instruments, while non-environmental taxes (non-dedicated) which can be converted to suit the objective of biodiversity conservation include Income Tax, Corporate Tax, Property Tax, Agricultural Income Tax, Excise Duty, Customs Duties and Value Added Tax. Of these non-dedicated taxes the first five are direct taxes and the remaining are indirect taxes. As far as non-dedicated taxes are concerned, they can work for environmental protection only in the event of incentives and rebates integrated into them for promoting environmentally sustainable activities or in the event of dis-incentives built into them for environmentally unsustainable activities.

Environmental taxes can partake on direct and indirect taxes. Pigouvian Taxes are direct taxes since they are directly imposed on environmental damages. The indirect environmental taxes are by contrast, levied on raw materials or other substances which contribute to environmental damages. Pigovian taxes have interesting impacts in terms of altered output and consumption behavior [11].

Pigouvian taxes is an innovative instrument in India in terms of biodiversity conservation in case we move towards a system of imposing a specific tax rate on emissions or effluents, particularly for airsheds and watersheds of biodiversity importance, which are subject to water or air pollution. The marginal social cost of pollution in such water or airsheds could be determined in terms of the opportunity cost of the use (consumptive or non-consumptive) values. Examples of this could be water pollution in Western Ghat watersheds, induced by industrial or commercial agricultural activities. A Pigovian tax based on a watershed or airshed approach may not entail amendment of the Water Cess Act.

4.2.6. Forest Development Taxes

The forest development tax is the only dedicated forest related tax in India. This is levied on forest products as defined under the Indian and State Forest Acts. The purpose of the levy is revenue consideration in the main, as it relates to forest produce harvested by the Forest Department. However, the term "forest produce" is extremely tricky. It may or may not include 'landraces' cultivated in jungles or forests by tribal's who enjoy settlement rights in forests. From the point of utilizing this tax in the cause of biodiversity conservation the term "forest produce" if expanded to include "genetic materials" can induce this instrument to confer positive biodiversity impacts.

4.3. Non-Dedicated Taxes

4.3.1. Restructuring Income Tax for Biodiversity Conservation

In India a large number of socially significant contributions are treated as deductions. An explicit deduction can be provided for contributions to charitable trusts that take up conservation of forests or Wildlife Reserves, or on donations made to environmental organizations working for conservation causes. Alternatively contributions made to the proposed National Biodiversity Fund of India in the category of investments may be made eligible for deductions.

4.3.2. Restructuring Corporate Tax for Biodiversity Conservation

The provisions of depreciation, if extended to assets such as 'natural areas' and 'conservation lands' which may be acquired by industrial undertakings, can have a salutary impact on biodiversity conservation [12(a, b), 13]. Investment tax allowances, commonly found in India countries provide for faster write-off of investment expenditure and could also partake of investment tax credits. However this does not extend to biodiversity use and biodiversity related investments such as investments in natural re-generation activities and contributions made to the Biodiversity Fund. Tax holidays can be extended to biodiversity conservation ventures, initiated by private sector enterprises in the fields of both *ex-situ* and *in-situ* conservation of biodiversity [4].

4.3.3. Agricultural Income Tax (AIT)

The agricultural income tax system in India follows a two track mode. The first track is that of composition whereby rate of agricultural income tax are related to holding size ranging from say, 15 to 50 acres. The second track of AIT is the filing of returns by holdings above 50 acres. At the same time plantation units with holding size of 15 to 50 acres can opt for filing or returns in lieu of composition system. The two track system is providing to be complex and cumbersome for the plantation units in addition to the fact that the levels of post-tax incomes are low to provide for adequate capital formation. Deductions under AIT are not linked to investments in conservation efforts. Rather they are associated with developmental activities such as productivity augmentation programmes which are often inimical to biodiversity conservation. In the coffee belt of India, large sections of non-planted areas under natural forests have been brought under coffee planting in the name of production and productivity improvement and developments. In future agricultural income tax should provide for deductions for conservation efforts made by plantations, including conservation of natural areas within plantations (3).

4.3.4. Property Tax

There are three basic forms of property taxation: (1) a tax based on the annual or rental value of the property; (2) a tax based on the capital value of the land and improvement; and (3) a tax based on the site (or land) value (which is essentially a type of capital value tax). Certain tax systems use a combination of methods. Under the annual value system, the tax is based on an estimate of the annual net rental value from the use of the property. Under the capital value system, the tax is based on the assessed value of land and improvements (3).

The old view of the incidence of the property tax was that the burden of the tax falls on land that is largely borne by owners of land since land is perfectly inelastic in supply. Since landowners tend to have higher income, this component of the tax that falls on structures, is borne by both the supply and demand sides of the market because structures are more elastic in supply. Since renters tend to have lower income, this component of the tax is less progressive in incidence or even regressive. Thus, under the old view, the property tax tends to be at best proportional and possibly regressive. The new view of property tax incidence is that the tax falls on all owners of capital through movements of capital from high-taxed sectors to low-taxed sectors. The efficiency of property tax depends upon the correct statement of the rental and capital value of property.

In U.S.A, Canada, Brazil and Guatemala, property tax exemptions are provided for natural reserves and conservation units in private lands. A similar provision should be made in India to utilize property taxes as biodiversity conservation instrument. Given the significance of such an exemption, it is desirable to improve the efficiency and effectiveness of this economic instrument. Alternatively one could also think of a higher incidence of property tax on private property situated near coastal areas or wildlife of forest reserves. The provisions of the Wildlife (Protection) Act, 1972 and India's Coastal Zone Regulations need to be employed for this purpose. Private property in and around Protected areas identified under the two legislations / regulations can be subject to higher incidence of property tax to secure positive biodiversity outcomes (3).

5. INDIRECT TAXES

5.1. Excise Duty

Excise taxes are taxes imposed on the consumption of selected goods, such as alcoholic beverages, tobacco products, and petroleum products. Goods selected for excise coverage typically exhibit one or more of the following characteristics: first, their production and sales are closely supervised by the government, that is, they are sumptuary goods or services; second, they are characterized by price-inelastic demand schedules; third, they have an income elasticity of demand greater than utility, that is, they are luxury goods or services; or fourth, their consumption is regarded by the government as lacking merit or as likely to cause negative externalities. Excise tax rates may be specific defined in either specific or ad valorem terms and they are usually much higher than the rates applied under the general sales tax of VAT.

We state the rationale for Excise Taxation as two-fold [14] *viz*:

Revenue Generation:

Large sales volumes, few producers, in elastic demand, easy definability, and a lack of close substitutes favor the use of exercises where revenue generation is the goal.

Excise as tool for improving vertical equity:

It is generally not advisable to expand the scope of excise taxes beyond sumptuary goods, fuel, and a few luxuries.

Specific versus ad valorem rates:

Conventional policy generally advocates that all excise duties be levied on an ad valorem basis rather than on a specific basis, because this protects the base of the tax from inflation. This is particularly important in developing countries because they are more likely to have high rates of inflation.

Specific rates may be better than ad valorem rates if tax administrative capacities are limited so that under valuation of domestic goods or imports is a common problem. Under a specific-rate excise tax, disputes over valuation do not arise, since the tax is not based on value but on physical units of the commodity. In addition, specific rates are the appropriate from of tax if the tax is intended to be externality correcting, that is, an ounce of alcohol should be taxed the same whether it is contained in high-quality spirits or not.

Ease of Administration:

Excise tax liability is typically incurred as the commodity leaves the factory gate. Revenue officers are typically stationed inside the production facilities, particularly in the case of breweries, distilleries, and cigarette factories, to maintain on site checks of production and shipments.

Making Excise Duties Biodiversity Friendly

The Raja Chelliah Committee on Tax Reforms (India, 1991) urged reforms of the excise duty system in India based on the guiding principle of adopting a broad-based and simple, but broadly neutral and moderate rate excise duty which would gradually move towards a full-fledged Value Added Tax (VAT).

In pursuance of the Chelliah Committee recommendations, the Union Government has, since the 1999-2000 Union Budget, moved towards a three tier excise duty structure. As per this structure, there is a merit rate of 8%, a central rate of 15% and non-merit rate of 24% excise duty to be levied on merit, moderate merit and non-merit goods. Items such as wood and non-wood based products particularly from natural forests should qualify to be in the non-merit category and attract the non-merit rate of 24%, in case, the excise duty is to convert itself into an effective economic instrument.

5.2. Sales Tax and Value Added Tax (VAT)

A comprehensive critique of Sales Tax and Value Added Tax (VAT) regimes is provided [15]. The Sales Tax system in India is administered by State Governments. There are two types of sales taxes: single-stage taxes and multi-stage taxes. A single-stage sales tax applies only at one stage of the production-distribution chain. For example, it may only apply to sales at the manufacturing, wholesale, or retail stage. In contrast, a multi-stage sales tax applies at several stages of the production-distribution chain. Examples of this type of tax may also include turnover taxes and the VAT. These taxes may apply to most sales of goods and services in the economy no matter at which stage of the production-distribution chain the sale is made. Alternatively, these taxes may include all sales up to a certain stage in the production-distribution chain such as manufacturing or wholesale stages. The sales tax and VAT also distinguish between merit goods and non-merit goods. Alcohol and Cigarettes are normally classified as non-merit goods, not goods which are based on forest production. A restructured sales tax or VAT should subject biological resources to a higher incidence of the tax in the larger interests of biodiversity conservation.

6. NEW ECONOMIC INCENTIVES/ FINANCIAL MECHANISMS

6.1. Price Incentives for Biodiversity Conservation

Price as an incentive for biodiversity conservation needs to be carefully pursued. An enhanced premium for sustainably produced industrial and agricultural products may be worked out. A process of certification of such products and produce should be put in place [16]. This could be extended to a large variety of 'spices' and other 'medicinal' products which are forests and natural area based. In addition, efforts should be made to dovetail the organic product standards laid down for agricultural commodities by the Ministry of Commerce and the APEDA with biodiversity relevant labelling standards in order to ensure premium prices for ecologically sustainable, agricultural products. A system of market creation for sustainable products should be attempted on the lines of a Brand Equity Fund for biodiversity friendly products.

6.2. Environment Performance Bonds in India

In addition to the dedicated and non-dedicated economic instruments described, there are a few other instruments such as the environment performance bonds which could also be employed for biodiversity conservation. Performance bonds which entail deposit of some of money with a Government agency prior to receiving a payment is a surety for anticipated damages to environment. The bond is returned to the subscriber if it is discovered that the site has not been damaged. Performance bonds can be tried out in respect of approved development projects under Forest (Conservation) Act and the Environment (Protection) Act. The bonds can be in relation to observance of the stipulations laid down by the Government of India under the two Acts, for execution of the project by the industry concerned. These bonds will be in addition to the obligations of the project authorities to deposit for compensatory afforestation and other environmental rehabilitation measures, since the focus of the bond will be on 'non-damages' stipulated in the clearance letters of the Government of India.

7. FUNDS

7.1. National Biodiversity Fund in India

The National Biodiversity Legislation of India has provided for the National Biodiversity Fund which will contain the proceeds of the license fee, benefit sharing royalties levied for access of genetic material or on commercial products developed on the basis of these materials, fees paid for collaborative research involving bio-materials and associated knowledge by bio-prospectors and research institutions, both national and international. Voluntary contributions to National Biodiversity Fund can be catalyzed by linking them to tax exemptions on such contributions [4].

7.2. National Green Mission of India

The Green India Mission which was announced by the Government of India in the year 2010 provides for a total outlay/cost of Rs 44,000 crores for 10 million ha of degraded lands over 10 years. These resources will be mobilized as 'additional' resource from India's planning process. Though not explicitly labeled as a fund, the 'earmarked' amount represents an additional financial source which aims to realize economies of scope by focusing in a balanced manner on both carbon and biodiversity outcomes [6].

CONCLUSION: ASSESSING PROSPECTS FOR INTRODUCING NOVEL MARKET BASED INSTRUMENTS

The Conference of Parties to the Convention on Biological Diversity in their tenth Session held at Aichi Nagoya, Japan adopted the Aichi Targets for biodiversity conservation. The Aichi Targets 20 in number, are capsuled into five strategic goals [17]. Target 3 requires harmful subsidies to be phased out and positive incentives to be introduced to promote biodiversity conservation, Target 20 of the Aichi Targets states that by '2020, at the latest, the mobilization of financial resources for effectively implementing the Strategic Plan for Biodiversity 2011-2020 from all sources, and in accordance with the consolidated and agreed process in the Strategy for Resource Mobilization, should increase substantially from the current levels.' The ideas expounded in the preceding paras are designed to fulfil the requirement of Target 3 as they attempt to seek redesign of existing dedicated and non dedicated tax and other fiscal instruments to provide positive incentives for biodiversity conservation in India.

The Aichi Targets do not explicitly provide for use of market based instruments such as Biodiversity Offsets, Habitat Banking and Payment for Ecosystems Services which have been analysed by scholars and development agencies [18-20]. There are many reasons why market based instruments of the type mentioned, cannot work efficiently in India given the present state of ownership over natural resources of critical biodiversity importance [21].

It is argued that biodiversity offsets and habitat banking can create incentives for conservation by increasing the costs of development and discouraging activities that reduce biodiversity. Since, developers must pay an additional fee for their activities in the form of compensation for residual biodiversity loss, this raises the costs of development pathways [18]. Secondly offsets are assumed to create demand for conservation land, thus providing an additional incentive to 'manage' land for conservation.

In practice, these assumptions may not hold in India for two reasons. In the first place, given the enormity of the poverty problem in rural areas and the importance of infrastructure development, the social costs of non development activities is high. This means that the costs of development are not high as compared to a non-state of development. Secondly, the significance of following the 'intensive margin' in agricultural operations is high, given low yields of agricultural crops and pressures of food insecurity. This creates an inherent problem in conserving agro biodiversity in India. Further, there are many physical restrictions in land use change that prevents even privately owned land based habitats from being left to the play of markets. Therefore unless there are strong fiscal incentives, by way of tax deductions for conservation and higher imposts against use of land based habitats for narrow lifestyle facilitating development purposes, the desirable outcomes may not be secured. Further the larger issue is of conserving traditional knowledge associated with biodiversity [22] and securing community action against undesirable land use changes that undermine biodiversity.

Those who actually enhance biodiversity-related goods should be benefitted for their efforts by those who actually benefit from biodiversity-related goods and services [23]. However the key issue is who pays and to what extent [18]. Valuation constraints and agency identification being constraints, the transaction costs of PES may be larger than conventionally thought of [24]. Finally since the State owns forests and wetlands, while communities possess these resources through de jure rights, operation of PES is fraught with difficulties.

There is a trend for conservationists to be votaries of command and control approaches towards conservation issues [25]. This paper does not follow this pattern. It seeks to provide adequate space for non regulatory measures that is based on the polluter pays principle. Economic Instruments form the first step towards creating the basic economic conditions for biodiversity conservation in India, given the institutional and development realities of the country. With progressive changes in property rights and regimes that permit markets to work for the cause of biodiversity conservation, a policy that favours introduction of market based instruments such as biodiversity off sets; habitat banking and Payment for Ecosystem Services can be effective, efficient and socially just.

ACKNOWLEDGEMENTS

Declared none.

CONFLICT OF INTEREST

The authors confirm that this chapter contents have no conflict of interest.

REFERENCES

[1] Perrings CA. Ecology, economics and ecological economics. Ambio 1995; 24(1): 60-64.

[2] www.teebweb.org/ TEEB. The Economics of Ecosystems and Biodiversity, An Interim Report, 2008.

[3] Damodaran A. 'Financing Adaptation and Conservation in India's semi-arid agro-ecosystems'. International Journal of Climate Change Strategies and Management, September 2012; 4(4): 386-403.

[4] Damodaran A. 'Economics and Policy Implications India's National Biodiversity Legislation', Economic and Political Weekly, Vol. 38, No. 49, 20, December 6-12 2003.

[5] Ministry of Environment and Forests: National Biodiversity Action Plan, Government of India (prepared by Arora, Sujata, Bhatt JR) November 2008.

[6] Damodaran A, Singha Rajdeep:Report on Assessment of Financial Resources required for Biodiversity Conservation in India for 2011-2020: Restructured paper. Indian Institute of Management Bangalore 2012.

[7] Bowles, Ian A. *et al.* Economic Incentives for Biodiversity Conservation: Encouraging Private Sector Support in Developing Countries, Discussion Draft, Conservation International Policy Papers, Washington, D.C. 1995.

[8] Goulder, Lawrence H. 'Environmental Taxation in a Second-best World'. In: Folmer Henk, Tietenberg Tom, Eds. The International Yearbook of Environmental and Resource Economics 1997/1998 – A Survey of Current Issues, Edward Elgar, Cheltenham, U. K. 1997; pp. 29-53.

[9] Bell, Gerwin.'User Charges and Environmental Laws'. In: Shome, Parthasarathi, Eds: Tax Policy Handbook, International Monetary Fund, Washington, D. C. 1995; pp. 104–108.
[10] OECD Promoting Sustainable Consumption, Good Practices in OECD Countries. OECD, Paris 2008, a.
[11] Krelove, Russell. Concepts of Tax Incidence. In: Shome, Parthasarathi Eds. Tax Policy Handbook. International Monetary Fund. Washington D. C. 1995; pp. 35 – 38.
[12] (a) King, John R. Corporate Income Tax. In: Shome, Parthasarathi Eds. Tax Policy Handbook. International Monetary Fund. Washington D. C. 1995; pp. 134 – 136.
(b) King, John R. Taxation of Capital Gains. In: Shome, Parthasarathi Eds. Tax Policy Handbook. International Monetary Fund. Washington D. C. 1995; pp. 155 – 158.
[13] Chua, Dale. Depreciation Schedules. In: Shome, Parthasarathi Eds. Tax Policy Handbook. International Monetary Fund, Washington D. C. 1995; pp. 136 –139.
[14] McCarten, William J., Stotsky Janet. Excise Taxes. In: Shome, Parthasarathi Eds. Tax Policy Handbook. International Monetary Fund. Washington. D. C. 1995; pp. 100 – 103.
[15] McMorran, Ronald T. A Comparison Between the Sales Tax and a VAT. In: Shome, Parthasarathi Eds. Tax Policy Handbook. International Monetary Fund. Washington D. C. 1995; pp. 82 – 85.
[16] Yadav A. Organic Food Production, Problems, Prospects and Opportunities. Ghaziabad: National Centre of Organic Farming, Ministry of Agriculture, Government of India 2011.
Alger, Keith. Scaling Up Economic Instruments for Biodiversity and Ecosystem Service Conservation in the Tropics, Economics and Conservation in the Tropics: A Strategic Dialogue January 31 – February 1 2008.
[17] Anonymous. Strategic Plan for Biodiversity 2011–2020 and the Aichi Targets:Living in Harmony with Nature, Secretariat of the Convention on Biological Diversity., accessed fromhttp:// www.cbd.int/doc/strategic-plan/2011-2020/Aichi-Targets-en.pdf 2011.
[18] OECD. Paying for Biodiversity: Enhancing the cost-effectiveness of payments for ecosystem services, New OECD Publication, Paris, October 2010.
[19] OECD. Report on Implementation of the 2004 Council Recommendation on the Use of Economic Instruments in Promoting the Conservation and Sustainable Use of Biodiversity,,ENV/EPOC/GSP/BIO(2008)1/FINAL, Paris 2008, b.
[20] ArildVatn, David N Barton, Henrik Lindhjem, SynneMovik, Irene Ring, Rui Santos: Can Markets Protect Biodiversity? An Evaluation of Different Financial Mechanisms, Noragric Report No.60, Department of International Environment and Development Studies, Noragric Norwegian University of Life Sciences, UMB, June 2011.
[21] Blom Martijn, Bergsma Geert, Korteland Marisa. Economic instruments for biodiversity: Setting up a Biodiversity Trading System in Europe. CE Delft Solutions for Environment, Economy and Technology, December 2008.
[22] Berkes, Fikret *et al.* 'Traditional Ecological Knowledge, Biodiversity, Resilience and Sustainability'. In: Perrings CA *et al*, Eds. Biodiversity Conservation. The Netherlands, Kluwer Academic Publishers; pp. 281-299.
[23] Wunscher T, Engel S, Wunder S. "Payments for environmental services in Costa Rica: increasing efficiency through spatial differentiation", Quarterly Journal of International Agriculture 2006; Vol. 45, No. 4.
[24] Alix-Garcia JM, Shapiro EN, Sims KRE. 'Forest Conservation and Slippage: Evidence from Mexico's National Payments for Ecosystem Services Program', Working Paper, Department of Agricultural and Applied Economics, University of Wisconsin, Madison 2010.
[25] Alger, Keith. Scaling Up Economic Instruments for Biodiversity and Ecosystem Service Conservation in the Tropics, Economics and Conservation in the Tropics: A Strategic Dialogue January 31 – February 1. 2008.

CHAPTER 16

Green Economics: It's Impact on Environment and Biodiversity

Jayamani C. V.*

School of Management Studies, CUSAT, Kochi, Kerala, India

Abstract: Environmental economics is a multi-disciplinary area of academic studies that focuses on the interface between developmental economics and natural ecosystems. Ecology and economics have originated from the same etymology. Environmental economics generally deals with the conservation of natural resources and its effective use in attaining sustainable development. Hence the issues related to the environment are inextricably interwoven to economic issues. Green economics has assumed greater importance in recent times. Greening the globe assumes more importance than the growth of the economy. The concerns of the corporate world on environment has achieved greater attention in the period of rapid population growth, lavish consumerism, urbanization, industrialization and advances in technology coupled with exploitation of resources. Unending consumption will not give unending satisfaction or happiness. The challenge before man is to secure happiness without increasing the level of exploitation, the scale of production and the rate of consumption.

Keywords: Biodiversity, bio-friendly living, consumerism, green economics, quality of life.

1. INTRODUCTION

Over the years man has achieved unbelievable progress with his intellect. But many of the activities undertaken by man had adversely affected the environment. Environmental pollution, acid rain, ozone depletion, green house effects, cancer and other health hazards are a few such global problems. Population explosion, compelling industrialization and consumerism has caused degradation of the earth. Life on planet earth is at great risk. In his mad craze for making his stay in this planet more comfortable and happy, man destroyed the natural balance of Mother Earth.

2. EVOLUTION OF ECONOMICS

The words ecology and economics stem from the same Greek root *okios* meaning

***Corresponding author Jayamani C.V.:** School of Management Studies, CUSAT, Kochi, Kerala, India; Mobile: 91 94470 59320; E-mail: cvjayamani@hotmail.com

Laladhas K.P., Oommen V. Oommen and Sudhakaran P.R. (Eds.)

house. Ecology is the study of relationships or interdependence between living organisms and their environment. Ecology deals with the household of nature, while economics deals with the household of man. Ecology studies harmony between nature and man whereas economy in general examines disharmony with man and nature.

Environment is a subset of human economy and the nature is very valuable to economics. Environment management gives more importance to the creation of natural wealth that is the planetary resources which are scarce in nature. The human activities of late caused much damage to the purity of the planetary resources like air, water and even the sunshine. He even polluted the *panchabhutas* (five natural elements like ether, air, fire, water and earth). Environmental management tries to balance the quality of environment and the quality of life of the people in the context of rapid growth and development [1].

The study of economics can be classified into three streams in the order of its evolution. They are pure economics, welfare economics and environmental economics. In pure economics earth has everything to satisfy everybody's need. Resources are plenty and the economists ensure its equitable distribution. Everybody is eco friendly in their attitude and there is no disturbance to nature by their activities. Welfare economics considers the increasing needs of the growing population. In a welfare society everybody's needs are to be satisfied. For that there should be maximum production using the available resources. Naturally it creates some disturbances in the distribution. The environmental economics believes that there should be sustainable production and sustainable consumption. We have to satisfy the need of the people without affecting the quality and capacity of the nature to produce the goods not only to this generation but for the future generation also (Table **1**).

Table 1: Evolution of Economics and its Effect and Impact on the Environment

Pure Economics	Resource plenty Equitable distribution	Earth has everything to satisfy the need of all
Welfare Economics	Maximum production Maximum consumption Equitable distribution	Satisfaction of needs and wants – welfare and wellbeing of the society
Environmental economics	Sustainable production sustainable consumption	Satisfaction of need without affecting the quality of environment

In pure economics resources are plenty and there is equitable distribution of those resources. Earth has everything to satisfy the need of all. The welfare economics ensures maximum production, maximum consumption and equitable distribution. The ultimate aim is to provide satisfaction of needs and wants. It ensures the welfare and wellbeing of the society. Environmental economics on the other hand keeps sustainable production and sustainable consumption. It provides satisfaction of need without affecting the quality of environment.

2.1. Environmental Economics

Environmental economics was once distinct from resource economics. But today both became associated with sustainability and is termed as radical green economics. It has paved the way for natural capitalism and environmental finance, which are considered to be the two distinct branches of environmental economics. In the word of Mike Nickerson, economy is three-fifth of ecology. Environmental Accounting, Environmental Capital, Environmental Auditing, Environmental Ethics have become sub-fields of the study.

2.2. Green Economics

It is an approach to economics in which the economy is considered to be a component of the natural world. It takes the widest possible view of stakeholders of a transaction to include the impact to nature, non-human spaces, the planet earth sciences, and the biosphere. Green economics includes environmental economics and ecological economics and is based on the equity principle and sustainable way of development. It is careful to include principles of social equity at the core of its concerns. Its emphasis is on the ecosystem as the right starting point for economics [2]. "Growthism" in economics disrupts and destroys growth of the life supporting capacity of the natural ecosystem. "Greeinism" has long been recognized as a strategy for the sustainable development of a country. The rural ecosystem is the right starting point of green economics which considers the environmental economics and ecological economics along with welfare economics. Greening the globe assumes more importance than the growth of the economy.

3. ECONOMY - ENVIRONMENT INTER LINKAGE

Economics is the system of deciding how to allocate limited resources that will be used to meet the needs and wants of a society. Economics is concerned about the equitable distribution of natural resources and its effective utilization for the welfare of the entire society. The ability of a country to buy most of what most people desire is an indication of its economic growth. It paved the way for consumerism. The sustainable economic growth of a country depends not only on the economic goals but also on the social and environmental goals, policies and strategies. Economics issues are closely linked to environmental concerns. For instance, over exploitation of renewable resources would undercut a country's economy. Concerns and issues related to the environment are inextricably linked to economic issues such as poverty. Poverty has been shown to drive people to indulge in destructive activities like felling of trees, exhausting agricultural land, and over exploitation of water like natural resources [3].

3.1. Different Stages of Economic History

In the beginning the relationship between man and nature was friendly. Man interacted with the nature with all respect and reverence in order to make a living in this beautiful earth. In the pre industrial period he started interfering in the matters of nature for better living. After the introduction of modern technology he started to indulge in nature's way by disturbing the ecological balance. Continuous human destruction started during post industrial period which inflicted irreparable damage to the fine fabrics of Mother Nature (Table **2**).

Table 2: Man - Nature Relationship through the Stages of Economic History

Period	**Nature of Action**	**Effect**
Pre-agri - Primitive 8000 BC	Human action in tune with nature	Interaction
Pre industrial. Civilized 1800 BC	Limited intervention	Interference
Modern, technological 2000 AD	Disturbance to ecological balance	Indulgence
Post industrial economy. 2200 AD	Continuous Human destruction	Irreparable damage

4. THE VALUE OF BIODIVERSITY

In order to maintain environmental quality and ecological balance, the environmental policies and programmes should be linked to economic growth and social needs. Environmental economics and environmental management assume greater importance in the study of rural development. It deals with the conservation of natural resources and it effective use for attaining sustainable development. It aims at the improved environmental performance which has a positive impact on human health and growth.

4.1. Environmental Impact

Every economic action can have some adverse effect on environment. Development with zero destruction is nearly impossible Every environmental change can have its adverse impact on the economy and *in situ*. Environment acts as a waste sink or as a partial recycling factor of human wastes during production and consumption. Environment is the source of energy and material resources. It is the supplier of amenity – educational and spiritual values to society. The rain forests, flora and fauna carry its own respective social welfare functions. Using the environment for one purpose – as a supplier of resources can reduce its ability to supply us with other services such as the ability to provide clean air. Overlapping the circles of services creates conflicts in resource use. Using a mountain region as a source of minerals means its amenity value is reduced. Using the river as a waste disposal unit means its amenity value is reduced (reduction in fish resources). Felling a forest for its timber reduces the electricity generating capacity of a dam, owing to soil erosion. It also reduces the amenity values since forest's inhabitants' animals and humans are displaced or destroyed [4].

4.2. Allocating Resources to Conflicting Demands

Economics is concerned with allocating resources to conflicting demands. The economic system particularly the market system works very poorly in allocating environmental resources. An imperfect specification of property rights results in ineffective setting of price which sends the wrong signals to producers, consumers and governments. As a result the individual benefits of preserving an environment understate the collective benefit of preservation. The economic activities of humanity have a profound impact on the natural environment in the form of rapid depletion of natural resources as well as through pollution.

The nature was generally viewed as blissful and bountiful. But modern man started abusing the environmental resources. Henceforth it became scarce resources. This paves the way for the transformation of environmental resources into environmental goods which are scarce. This causes the imbalance in demand and supply. Most of the environmental goods become short in supply against the multiplication of demand. The reduction in both quality and quantity of natural resources is attributed to the affluence of people that accompanies economic growth and developments. An improvement in the quality of life and standard of living will improve the quality consciousness of environment.

5. CORPORATE WORLD AND ENVIRONMENTAL CONCERNS

Humanity is on the cross roads to find solutions to the environmental problems mostly created by the lifestyle changes of the modern people at local, national and global level. The top level engineers and executives should design, plan and implement strategies with a mentality of corporate environmental responsibility. Their actions should address the environmental issues along with the economic viability and technical feasibility. Environmental concern of the corporate world assumes greater importance in the period of rapid population growth, lavish consumerism, urbanization, industrialization and advance in technology coupled with exploitation of resources.

Bio friendly living and big business should be companions in their endeavour to protect environment. Unending consumption will not give unending satisfaction or happiness. The challenge before man is to secure happiness without increasing the rate of consumption. The aggressive advertisement and conspicuous consumption in the economy put imbalance in the equity principle of environment. "A cultural policy of controlling consumption will have to be made. Awareness should be created among the young and the old alike that happiness comes from discovering one's inner self rather than from mindless increase in consumption" [5].

6. EXCHANGE OF MONEY AND ENVIRONMENTAL BEHAVIOUR

Money is important in economics. In economics we exchange money for goods and services. Exchange of money can express our environmental behaviour, which is expected to be eco-friendly. An eco-friendly man will be conscious about the quality of the environment in his spending. He will exchange money for eco friendly products like cotton shirts and herbal drinks, ethnic food *etc*. He will give preference for bicycles or battery operated bikes, carbon neutral computers, low radiation cell phones, recycled papers and environmentally responsible products. He prefers to dine at a most eco- friendly restaurant, stay at carbon neutral hotel and go for a carbon neutral ride [6].

CONCLUSION

What we need today is several Schumachers to propagate the beauty of smallness and about green economics. We need thousands of Fakuokas to experiment with this earth in an eco friendly manner (Masnavao Fukuoka – the author of One Straw Revolution). Laurie Bakers may incarnate here thousand times teaching us about low cost, eco friendly, high quality green homes. We need real Gandhians who can teach us about the need for moderation in eating fasting and simple living which is eco friendly. We have to maintain the repertoire of grandma and her conventional wisdom that a locally produced organic food is thousand times better than the food which travelled thousand miles across the seven seas. We should realize that a crowded shopping mall is environmentally inferior to our small village markets.

ACKNOWLEDGEMENTS

Declared none.

CONFLICT OF INTEREST

The authors confirm that this chapter contents have no conflict of interest.

REFERENCES

[1] Sharma. Environment, Man and Nature. Reliance Publishing House, Delhi. 1989.

[2] Kurian Joseph, Nagendran R. Essentials of Environmental Studies. Pearson Education (Singapore) Pvt. Ltd. New Delhi. 2005.

[3] Mohandas M. Ecologically Sustainable Development: Concepts and Issues. Kerala Agricultural University, Thrissur, Kerala. 1993.

[4] Muraleevallabhan TV. Dimensions of Sustainable Economic Development, UNMA Publications, Nooranad. 2005.

[5] Jayamani CV, Vasanthagopal R. Environmental Management from Vedic to Modern time. New Century Publication, New Delhi. 2013.

[6] Jayamani CV. Kerala Lifestyle for Modern Youth. ICKS, University of Kerala, Thiruvananthapuram. 2007.

Chapter 17: Biodiversity Valuation of Marine Ecosystem of the South-West Coast of India with Special Reference to Kerala Coast

CHAPTER 17

Biodiversity Valuation of Marine Ecosystem of the South-West Coast of India with Special Reference to Kerala Coast

Joshi K.K.*, Narayanakumar R., Ranimary George, Molly Varghese, Jasmine S., Srinath K.R., Saravanan S., Somy Kuriakose, Geetha Antony, Geetha P.M. and Gimy George

Marine Biodiversity Division, Central Marine Fisheries Research Institute, Kochi, Kerala, India

Abstract: Measuring biodiversity has been a challenge in valuing and developing quantitative indicators for its management and conservation. The faunal and floral species diversity and richness makes the south west of India one of the hot spots of biodiversity as well as an area of important multispecies fisheries. Among the 118 species of sea weeds reported, majority belong to the family Rhodomelaceae (13), followed by Caulerpaceae (10), Cladophoraceae (6), Ulvaceae (5), Dictyotaceae (5). A total of 730 molluscan species were enlisted from southwest coast which include 515 species of gastropods, 171 species of bivalves, 28 species of cephalopods and 14 species of scaphopods. Crustacean diversity is shown by the presence of 218 species including 152 species of crabs, 57 species of shrimps and 9 species of lobsters. 950 fish species occur along the coast of which 800 species are landed in commercial fishery. Total fish landings from the area were about 1.187 million tonnes. Marine fish landings from Kerala was 7.43 lakh tonnes of which more than 50% was contributed by oil sardine, mackerel, threadfin breams, carangids and penaeid prawns. Out of the 21871 fishing crafts along Kerala coast 3768 are trawlers, 60 purse seiners, 460 gillnetters and 495 ring seiners. The value estimated for the ecosystem services and natural capital of Kerala coast is of US $ 1660-1930 billion per year from an area of 260101 km^2 which includes brackish water, estuaries and Open Ocean.

Keywords: Biodiversity indicators, biodiversity valuation, bio-resources, Kerala coast, southwest coast.

1. INTRODUCTION

Coastal ecosystems are the most productive areas in nature but they are highly threatened systems also. They provide more services to human beings than any

***Corresponding author Joshi K. K.:** Marine Biodiversity Division, Central Marine Fisheries Research Institute, Kochi, Kerala, India; Tel: 0091-0484-2394795; Fax: 0091-0484-2394909; E-mail: joshyguru@gmail.com

Laladhas K.P., Oommen V. Oommen and Sudhakaran P.R. (Eds.)

other systems but face the most rapid environmental deterioration. The south-west coast region of India extends from 8° N to 15 °30' N with a coastal line of 995 km in three maritime states, Kerala, Karnataka and Goa and has got a shelf area of 75400 km^2. The peculiar wind and water circulation during south-west and north-east monsoon makes this a unique ecosystem as compared to the similar areas of same latitude. Most important biodiversity indicators of the coast are mud banks and monsoon fishery which provides livelihood to about 6.0 lakh people.

In general taxonomic trees form a connecting link between economic and ecological measurement of diversity whereas in a situation where nothing is available, the taxonomic data and information have value. As indicated in the Convention of Biodiversity (CBD) the economic valuation of biodiversity and biological resources is an important tool for well-targeted and calibrated economic incentive measures. The project on biodiversity valuation of marine ecosystem was taken up with the objectives to understand the ecologic and economic value of the diversity of marine fishery resources and related activities carried out in the southwest coast of India and to investigate the economic valuation of the important groups of the organisms of south-west coast of India.

2. BIODIVERSITY VALUATION

Preliminary valuation of different organism and data on the biodiversity information and economic values of fish and other marine organisms of southwest coast of India were collected during the period 2011-2012. Species list of different groups like phytoplankton, zooplankton, coelentrates, annelid, crustaceans, echinodermata, sponges, gorgonids, molluscs, fishes, mangroves, reptiles, birds, and marine mammals were prepared from the present study and additional information from the past literature. The fishery data of the different organisms has generated the Fishery Resource Assessment Division (FRAD) data base of Central Marine Fisheries Research Institute (CMFRI).

2.1. Phytoplankton

A list of 468 species belonging to 154 genera, 89 families and 55 orders were recorded. Among the 55 orders, diatoms (Class Bacillariophyceae) dominated with 27 orders, followed by Class Dinophyceae with 10 orders, Class Coscinodiscophyceae with 8 orders, Class Mediophyceae with 7 orders and other Classes formed by the rest of the orders. Family wise analysis shows that 39 families belong to the Class Bacillariophyceae, 23 families to the Class Dinophyceae, 14 families to the Class Coscinodiscophyceae, 11 families to the Class Mediophyceae and the rest of the families recorded belong to the Classes Cyanophyceae, Dictyochophyceae, Trebouxiophyceae, Ulvophyceae, Chlorophyceae, Siphonocladophyceae, Raphidophyceae, Euglenophyceae, Coccolithophyceae and Chlorodendrophyceae.

2.2. Zooplankton

Zooplankton group contains wide variety of species from about nine different phyla such as protozoa, coelentrata, ctenophore, rotifers, annelid, polychaeta, chaetognatha, arthropoda, mollusca and chordate. About 251 species of organisms belonging to 170 genera, 84 families and 36 orders of different phylum were recorded from this coast. The zooplankton group form one of the important components in the larval feed of many of the marine organism. The survival and success of the organisms depend on the availability of the zooplankton. The species diversity and species richness of the south west coast makes it one of the diverse regions [1-3].

2.3. Coelentrates

The coral reefs of the southwest coast of India are rich in biodiversity. The coral fauna includes the subclass Hexacorallia (15 families) and among this, the family Acroporidae (55 spp.) dominated followed by Faviidae (34 spp.), Fungiidae (19 spp.), Poriitidae (18 spp.), Agaricidae (13 spp.) Pocilloporidae (10 spp.) and Mussidae (10 spp.). Large scale exploitation of coral reefs for industrial purpose, their fauna and flora for multipurpose uses and dredging result in the consequent mass mortality of corals.

2.4. Annelida

Nematode diversity along the Kerala coast was highlighted by several studies [4-6]. A total of 154 species of nematods have been reported from West coast of India [7]. The studies indicated that changes occur in community structure due to pollution, organic enrichment and anthropogenic activities. Also the seasonal rains during the monsoon and post-monsoon periods influence the water quality and resulted in species variation.

2.5. Crustaceans

The present study revealed the presence of 218 species of crustaceans which included 152 species of crabs, 57 species of shrimps and 9 species of lobsters along the coast. Shrimp diversity included 9 families, 22 genera and 57 species while crabs showed maximum diversity of 20 families, 86 genera and 152 species. The species lists of crustacean were documented in the past on shrimps lobsters and crabs [8].

The prawn resources of Ashtamudi Lake comprised of nine species of penaeids, five species of Carideans, one species of sergestid [9]. The occurrence of large concentration of juveniles established the fact that backwaters and estuaries are the nursery grounds of the prawns. The deterioration of the estuaries and backwater system caused negative effects on prawn fishery and associated trophic groups. Destructive fishing by minitrawls along the Kerala coast on the juveniles of *Parapeneopsis stylifera* is an example of recruitment overfishing and consequential conservation problems [10]. *Penaeus indicus* are available appreciably in the estuaries while *Metapenaeus dobsoni* are abundant in areas of Kasragod, Madakara, Ponnani, Puthuponnani, Chetwai, Azheekod, Ramanhturuth, Manajanakkad, Panavalli and Kumbalangi [11].

2.6. Echinodermata

The present study revealed the presence of 112 species of echinoderms belonging to 70 genera, 35 families and 16 orders along the coast. The sea cucumbers, *Holothuria artra, H. nobilis* and *Actinopyga* are the important species occurring along the South west coast of India. As a conservation measure all the Holothurians are put under the Wildlife Protection Act of Ministry of Environment and Forests.

2.7. Sponges

Sponge fauna of Lakshadweep and Kerala coast, revealed a total of 91 species recorded from this area. The class Demosponginae was represented by 45 families from this area. Among this, the family Clionaidae dominated with six species, followed by Spirastrellidae (4), Halichindridae (4) Tethyidae, Spongidae, Suberitidae and Agelasidae with three species each. The families Timeidae, Axinelluidae, Microcionidae, Thorectidae, Myxillidae, Ancorinidae and Dysideidae were represented by two species and the families Callyspongidae, Anocorinidae, Phloedictyodae, Mycalidae, Raspaillidae, Samidae, Spongiidae, Tetillidae, Suberitidae, Dictyonellidae, Halichondridae, Axinellidae, Amphilecitidae, Lophonidae, Acamidae, Darwinellidae, Micricionidae, Chondrillidae, Raspallidae, Plakinidae, Niphatidae, Geodildae, Thorectidae, Iotrochotidae, Ancorinidae, Theonellidae, Desmanthidae, Placospongidae, Pseudoceratinidae and Microcionidae were represented by single species only.

2.8. Gorgonids

Among the 34 species of gorgonids reported, majority belongs to the family Plexauridae (10), followed by Ellisellidae (7), Melithacidae (2) Acanthogorgidae (2) and Subergorgiidae (2). The families Claverlariidae, Nephtheidae, Isididae, Anthothelidae and Gorgoniidae were represented by single species only. The important species of gorgonids distributed along the Kerala coast are *Echinomuricea indica, Heterogerogia flabellum, Echinogerogia complexa* and *Gorgonella umbraculum.* They are biomedically important resource and large quantities have been exploited for export purposes.

2.9. Mollusca

A total of 730 molluscan species were enlisted from southwest coast which include 515 species of gastropods (75 families, 196 genera) 171 species of bivalves (35 families, 91 genera) 28 species of cephalopods (12 genera) and 14 species of scaphopods (5 genera). Crustacean diversity is shown by the presence of 218 species of which 152 species of crabs, 57 species of shrimps (20 families, 86 genera) and 9 species of lobsters (3 families, 3 genera). Commercially important bivalves are *Villorita cyprinoides, Paphia malabarica, Meretrix casta, Mercia opima, Crassostrea madrasensis, Saccostrea cucullata* which contribute about 59000 tonnes annually. The clam fishery was about 35000 tonnes annually from Vembanad and Ashtamudi lake areas. Edible oysters occur in all the estuaries and grow abundantly in Ashtamudi, Vembanad, Mahi, Valapatnam and Neelesswaram backwaters and estuaries. The clam beds are seen in Ashtamudi and Vembanad lakes and the high saline Malabar area are occupied by mussel beds. The mussels of species *Perna viridis* (green mussel) and *Perna indica* (brown mussel) contribute an annual landing of about 8000 tonnes. Green mussel has an abundant distribution whereas brown mussel has limited distribution from Quilon to Kanyakumari. The Sacred Chank (*Turbinella pyrum*) is important for the Hindus and are used in worship and for manufacturing bangles by the ladies in Bengal.

2.10. Fishes

Fish species occurring along the coast were about 950 species (175 families, 42 orders) of which 200 species are used in the commercial fishery. Commercial fishery scenario of Kerala contains variety of gears including trawl nets, drift gillnetters, bottom set gillnets, ring seines, hooks and lines boat seines. In the gillnet sector a variety of modification was made to suit to catch varieties of species. The marine fishery of Kerala is dominated by trawls. The trawl fishery contributes to about 72% of the mechanized sector. The trawl landings are concentrated at Quilon, Ernakulam and Calicut districts due to the availability of harbour facilities. The trawlers conduct targeted fishery for prawns off Kanyakumari in the South and Beypore (Calicut) in the north during the period November to March. The trawlers from Neendakara area generally do the fishing in the grounds lying between Kayamkulam and Anjuthengu and Cape Comorin region. The most productive area in the south-west coast is the Quilon Bank in the depth range of 275-375 m off Quilon-Alappuzha with an area of 3300 sq. km. The Wedge Bank of Cape Comorin is also an important fishing area. The north-west of Cochin to Chettuva is known as Perch grounds with a depth of 7-200 m [12].

The highest species-diversity was observed in the family Carangidae (28 spp.) followed by Serranidae (17), Engraulidae (12), Leioganthidae (12), Lutjanidae (10), Sciaenidae (9), Nemipteridae (8), Bothidae (7) and Soleidae (7), Cynoglossidae (6), Gerreidae (6), Mullidae (6), Platycepahlidae (6) and Scombridae (6), Dasyatidae (5) Rhinobatidae (5), Callionymidae (5) and Clupeidae (5) [13]. Recent studies revealed that three species *Scomberomorus commerson, Sardinells longiceps* and *Saurida* species are abundant and *Cynoglossus* sp., *Auxis* sp. and *Nemipterus* sp. are less abundant [14].

2.11. Mangroves

The mangrove ecosystems are of diverse assemblage of salt tolerant intertidal flora and fauna and are a major nursery ground for the larvae and juveniles of many commercially important species of prawns and fishes. The present study revealed the presence of 33 species of mangroves belonging to 26 genera and 18 families. The mangroves also face problem of mass scale denudation of mangrove wood, heavy siltation and conversion to prawn farms [10]. Of the 17 sq. km mangrove area, 36% is either completely degraded or is degrading. The destruction of mangroves is usually proportional to human population density. The major reasons for destruction of Mangroves are industrial pollution, oil spills, storms, dredging for landfalls and building ports, industrial estates, housing estates for human habitation [15].

2.12. Reptiles

Five species of sea turtles, *Dermochelys coriacea, Eretmochelys imbricata, Chelonia mydas, Lepidochelys olivacea* and *Caretta caretta* are endangered and vulnerable. Sea snake diversity of Southwest coast of India includes about 25 species.

2.13. Birds

Bird diversity along the southwest coast is very well documented from different estuaries, backwaters and Kole wetlands. Varieties of birds are seen along the mangrove areas, estuaries, backwaters, kole wetlands of Kerala [16]. About 76 species of birds are known to occur in Kerala in association with mangroves [16 - 18].

2.14. Marine Mammals

The marine mammals belonging to the whales, dolphins, porpoises are rare and endangered. A total of about 10 species of whales have been reported from southwest coast. They are *Indopacetus pacificus* (Longman's beaked whale), *Balaenoptera borealis, B. musculus, B. acutorostrata, Pseudocra crassidens, Pysester macrocephalus, P. catodon, Ziphius carvirostris* and *Balaenoptera sp.* A total of five species of dolphins were recorded including *Stenella longirostris* (Spinner dolphin), *Sousa chinensis* (Humpback dolphin), *Delphinus dephis* (Common dolphin), *Tursiops truncates* (Bottle nose dolphin) and Risso's dolphin.

3. BIODIVERSITY INDICATORS

The important components of biodiversity along the Kerala coast are coastal fisheries, mud bank ecosystems, productive rocky bottoms, Malabar upwelling area, Island and Coral ecosystems, mussel beds, and traditional coastal mariculture area. The total area of brackish water area has reduced from 4226 km^2 to 652 km^2 at present due to reclamation for various purposes [19]. Estuaries are highly productive and form nursery grounds for a variety of commercially important species. The backwaters are used for transport of toxic materials by barges, public transport, waste disposal, mariculture, recreation, reclamation for agriculture and residential development. Domestic and industrial effluents discharged from the urban habitations and industries have led to instances of fish kill, degradation of benthic biota and a general decline in the natural carrying capacity of the system.

3.1. Mud Banks

The mud banks (Chakara) are 1 to 3 m thick in calm, turbid near shore waters of 2-5 km along the coast with a width of 1.5 to 4 km which appear along with southwest monsoon. Mud banks are known to exist from time immemorial and has always been a matter of curiosity and mystery. Mud banks can be divided into four depending on the source of the mud. Accordingly mud can be formed by subterranean mud, by the aggregation of coastal mud, by the sediments discharged from rivers and estuaries and by the accumulation of mud from dredging operations. The results obtained so far are not final and further refinement of the hypothesis and concepts are necessary. There are several anthropogenic activities negatively affecting the mud bank formation along the Kerala coast. Several past studies indicated the importance of mud banks to commercial fishermen of Kerala during the monsoon period [20]. But the change in the monsoon pattern reduced flow from rivers due to several reasons; changes in the hydrographic pattern are decision factors in the formation of mud banks. Over the years, it has been observed that duration of mud banks, areas of occurrences and intensity have been declining due to the factors mentioned above.

3.2. Pokkali Fields and Prawn Farming

In Kerala, the total brackish water resources including the lower reaches of rivers, the brackish water lakes, the backwaters and the adjacent low-lying fields and mangrove swamps are estimated at about 2, 43,000 hectares. A traditional system of farming, Pokkali rice cultivation using highly salt resistant 'Pokkali rice' is another important uniqueness of Kerala coast. The microheterotrophs in the pokkali fields are represented by protozoans, metazoans and bacteria. The interaction between phytoplankton and protozoan are the major pathways in the cycling of nutrients in the system. Recently, the Pokkali rice got GI (Geographical Indication) status which enhanced marketing and generated better returns. However, the degradation and loss of habitat, loss of associated species in the pokkali fields and other socioeconomics issues have to be addressed immediately to conserve this important traditional farming system.

Pokkali rice cultivation is characteristic of the coastal saline soils of Kerala along the districts of Ernakulam, Trichur and Alappuzha. Within this Pokkali rice high genetic diversity exists and varieties like Pallipuram Pokkali, Kuzhuppally Pokkali, Vettikkal Pokkali are grown in different regions [21]. In pokkali system one rice crop is followed by fish or prawn which provides substantial income to the farmer.

4. THREATS TO BIODIVERSITY

Inshore areas of coast are under severe stress due to increasing fishing activities which resulted in the exploitation of juveniles, damage to benthic fauna, and stock depletion of some of the resources. Most of the marine mammals, turtles, catfishes, flatheads, whale shark, unicorn cods, flying fish, whitefish, lobsters and sharks became threatened and vulnerable [10]. The major industries located along the estuarine and coastal areas discharge waste materials into the coastal zone in the form of chemical toxic materials [19]. Studies indicated that some populations emerged with a likelihood of extinction and have peculiar characters such as long life span, slow growth, low reproductive effect, large size at sexual maturity, poor regeneration, short dispersal, near shore occurrence, narrow depth range, small geographical range, high vulnerability and small population size.

5. BIODIVERSITY VALUATION

Total fish landings from the area were about 1.187 million tonnes (31% of all India landings). Marine fish landings from Kerala was 7.43 lakh tonnes of which more than 50% was contributed by oil sardine, mackerel, threadfin breams, carangids and penaeid prawns. Total number of fishermen population was 6, 10,165 of which 1,30,922 were engaged full time and 10582 as part time. Number of fishing crafts along Kerala coast were 21871 of which 3768 trawlers, 60 purse seiners, 460 gillnetters and 495 ring seiners form the capital investment. The preliminary economic valuation of fishery resources was done collecting data on direct value, indirect value and other values. The value estimated for the ecosystem services and natural capital of Kerala coast is US $ 1660-1930 billion per year from an area of 260101 km^2 which includes brackish water, estuaries and open ocean. Values showed an increasing order from open ocean, *Exclusive Economic Zone (*EEZ), continental shelf, estuaries and brackish water area. Thus biological diversity plays an important role for essential ecosystem services and goods particularly in the form of fish and fishery products.

CONCLUSION

The study provides much needed information on the species diversity, richness and biodiversity value of resources in relation to their ecosystem. An important lacuna in the knowledge is the lack of information on the various services provided by the system and benefits offered by Kerala coast on a locality basis. Considering the above gaps in the information and importance of biodiversity valuation of marine ecosystems, CMFRI have studied the total economic value of the biodiversity of the selected groups of important resources in the ecosystem so that necessary policy interventions are evolved for long term and sustainable management of the bio-resources of the marine sector.

The Ecosystem Based Fisheries Management (EBFM) approach develops models of fisheries dynamics, trophic relations and ecosystem interactions. The model compares the fishing ecosystem with the natural ecosystem more reasonably than any other models. By incorporating trophic dynamics into EBFM models a change from traditional population based analysis to more realistic approaches was achieved. Future fishery management approaches will be mainly based on EBFM and CBFM (Community Based Fisheries Management) rather than traditional models for sustainable utilization of the resources for the present and the generations to come.

ACKNOWLEDGEMENTS

Declared none.

CONFLICT OF INTEREST

The authors confirm that this chapter contents have no conflict of interest.

REFERENCES

[1] Silas EG, Pillai PP. Dynamics of zooplankton in a tropical estuary (Cochin backwaters) with a review on the plankton fauna of the environment. Bull. Dept. Mar. Sci., Univ. Cochin 1975; 7(2): 329-355.

[2] Naomi TS, Ansy Mathew, George JP, Giri S, Kaliamoorty M. Zooplankton fauna. CMFRI Special Publication Mangrove Ecosystems: A manual for the assessment of biodiversity 2005; 83: 36-82.

[3] Mathew KJ, Gisha Sivan, Krishnakumar PK, Somy Kuriakose. Euphasids of the west coast of India. CMFRI Special Publication 2003; 78: 1-155.

[4] Damodaran R. Studies on the benthos of the mudbank regions of Kerala coast. Ph.D. Thesis, University of Cochin, Kerala India 1973.

[5] Sajan S, Damodaran R. On the vertical distribution of marine nematodes from the Shelf sediments off west coast of India with focus on their community structure. J. Mar. Biol. Ass. India 2005; 47: 133-138.

[6] Anilakumary KS. Diversity of meiobenthic nematodes in the Poonthura estuary (Southwest of India). J. Mar. Biol. Ass. India 2008; 50(1): 23-28.

[7] Sajan S, Joydas TV, Damodaran R. Meiofuana of the western continental shelf of India, Arabian Sea. East. Coast. Shelf. Sci. 2010; 86(4): 665-674.

[8] Mohammed KH, Vedvyasa Rao P. Estuarine phase in the life history of commercial prawns of the west coast of India. J. Mar. Biol. Ass. India 1971; 13(2): 149-161.

[9] Suseelan C, Kathirvel M. A study on the prawns of Ashtamudi backwaters in Kerala with special reference to penaeids. Indian J. Fish. 1982; 29(1&2): 71-78.

[10] Devaraj M. Coastal biodiversity-conservation and sustainable management. In: Menon NG, Pillai CSG, Eds. Marine Biodiversity conservation and management. Cochin, CMFRI 1996; pp. 193-205.

[11] Rao PV. Penaeid prawn seed resources in the estuaries and backwaters of Karnataka and Kerala. Mar. Fish. Infor. Ser. T& E Ser. 1980; 20: 9-11.

[12] Ammini PL, Srinivasan J, Ramani K, Beena MR, Seynudeen MB. Marine fisheries in Kerala - an overview. Mar. Fish. Infor. Ser. T& E Ser. 2010; 204: 1-9.

[13] Naomi TS, Rani Mary Geroge, Miriam Paul Sreeram, Sanil NK, Balachandran K, Thomas VJ, Geetha PM Finfish diversity in the trawl fisheries of southern Kerala. Mar. Fish. Infor. Ser. T & E Ser. 2011; 207: 11-21.

[14] Mohammmed KS, Sathianandan TV, Zacharia PU, Asokan PK, Krishnakumar PK, Abdurahiman KP, Veena Shettigar, Durgekar RN. Depleted and collapsed Marine Fish stocks along the Southwest coast of India- A simple criterion to assess the status. In: Meenakumari B, Boopendranath MR, Edwin

Leela, Sankar TV, Gopal Nikhita, Ninan George, Eds. Coastal Fishery Resources of India, conservation and sustainable utilization. Society of Fisheries Technologists, Cochin 2010; pp. 67-76.

[15] Satheeshkumar PV, Manjusha V, Pillai NGK. Conservation of mangrove forest cover in Kochi coast. Current Science, 2011; 101: 1400.

[16] Jayson EA. Structure, composition and conservation of birds in Mangalavanam mangroves, Cochin, Kerala. Zoos print Journal 2000; 16(5): 471-478.

[17] Kurup DN. Ecology of the birds of Bharthapuzha estuary and survey of the coastal wetlands of Kerala. Final report submitted to Kerala Forest Department, Trivandrum 1996.

[18] Prasanth Narayanan S, Thomas AP, Sreekumar B. Ornithofauna and its conservation in the Kuttanad wetlands, southern portion of Vembanad-Kole Ramsar site, India. Journal of Threatened Taxa 2011; 3(4): 1663-1676.

[19] Devaraj M, Menon NG, Pillai VK, Balachandran K. Coastal zone convergence: Kerala in relation to national and international situations. Mar. Fish. Infor. Ser. T & E Ser. 1999; 162: 1-22.

[20] Nair PVR, Gopinathan CP, Balachandran VK, Mathew KJ, Regunathan A, Rao DS, Murthy AVS. Ecology of Mudbanks - Phytoplankton productivity in Alleppey Mud bank. CMFRI Bulletin 1984; 31: 28-35.

[21] Leenakumary S. Rice varieties Kerala's special. Kerala calling, October 2007; pp. 16-18.

Chapter 18: Survey and Assessment of the Status, Issues Threats to Marine Turtles Along the Kerala Coast

CHAPTER 18

Survey and Assessment of the Status, Issues & Threats to Marine Turtles Along the Kerala Coast

Renjan Mathew Varghese*

WWF-India, Kerala State Office, Vanchiyoor P.O, Thiruvananthapuram-695 035, Kerala, India

Abstract: Mainly seven species of turtles are reported across the world, out of which five are found in India, and four - Olive Ridley (most common), Green Turtle, Leatherback and Hawksbill have been sighted from the 9 coastal districts of Kerala. This study identifies the major beaches along the Kerala coast where turtles have been sighted, their nesting areas, and threats to their survival and current efforts taken to conserve them. Field surveys, interacting with the coastal people and liaising with local conservation groups were conducted. The coastal belt of Kasargod has the largest density of marine turtles followed by Kozhikode and Thrissur. The year 2010 saw a 70% surge in turtle landings in these regions over the previous year due to the continual efforts of local conservation groups – Naithal. Many turtles trapped in fishing nets die and sometimes the fishermen kill the live ones for meat or as revenge for destroying the nets. Beach erosion and coastal afforestation and burgeoning coastal development in the form of ports, sea walls, fishing harbours and tourism are major threats to turtle habitats and nesting sites. The study recommends awareness campaigns for turtle conservation, banning illegal sale of meat and eggs, initiating Coastal Eco Development Committees, diligent afforestation, shoreline protection using geo-textiling, curbing contamination of coastal area by waste dumping and industrial drains and co-ordination between the concerned authorities for better vigilance of beach sand mining, illegal trawling, *etc.*

Keywords: Beach erosion, Casuarina plantations, coastal afforestation, coastal biodiversity, coastal development, coastal pollution, conservation groups, Green turtle, habitat loss, Hawksbill, illegal market, Kerala beaches, Leatherback, marine turtles, Olive Ridley, polluting activities, sand mining, trawl nets, turtle conservation, turtle meat.

1. INTRODUCTION

The State of Kerala lies between the latitudes 8° 17' 30" and 12° 47' 40" N and longitudes 74° 27' 47" and 77° 37' 12" E. Kerala has a coastline of about 590kms.

***Corresponding author Renjan Mathew Varghese:** WWF-India, Kerala State Office, Vanchiyoor P.O, Thiruvananthapuram-695 035, Kerala, India; Tel: 91-0471-2302265; E-mail: renjanmv@wwfindia.net

Laladhas K.P., Oommen V. Oommen and Sudhakaran P.R. (Eds.)

Out of the 14 districts in the State, 9 of the districts have the coastline running through it. The total area of the coastal districts comes to about 22,418 sq.kms.

Table1: The district wise distribution of coastline of Kerala

l. No.	District	Length of Coastline	
		Length (in Km)	Percentage of total
1.	Thiruvananthapuram	78	13.20
2.	Kollam	37	6.30
3.	Alappuzha	82	13.90
4.	Ernakulam	46	7.80
5.	Thrissur	54	9.20
6.	Malappuram	70	11.80
7.	Kozhikode	71	12.00
8.	Kannur	82	14.00
9.	Kasargode	70	11.80
	Total	590	100.00

Source: ACCESS Knowledge Series [1].

Kerala is considered as one of the states having the highest population density and within Kerala it is the coastal areas which have the highest population and tenement density. The coastal population of Kerala is around 93,97,625 with a density of 2022 persons/sq.km. This makes the coastal areas of the State more vulnerable. The coastal ecosystem of Kerala is abundant with various ecosystems and its inhabiting species.

"The coastal zone in Kerala is the low land fringing the sea extending over 590 kms, with a height of less than 8m from the Mean Sea Level (MSL), covering about 15% of the State's total area of 38,863 sq.km" [2]. "The vast low lying area consists of beach dunes, ancient beach ridges, barrier flats, coastal alluvial plains, flood plains, river terraces, marshes and lagoons " [2]. "The maximum width is found in the Alappuzha and Aluva - Kaladi regions" [2]. "This region is characterised by the presence of beach drive ridges especially in the Alappuzha - Cherthala regions" [2]. "Further the coastal plain is seen to extend between numerous rocky ridges along the coast." [2]

More than 260km of the coast is subjected to erosion. Shore protection measures such as seawalls, seawall & groin assembly and groins have been taken up along the coast. Another major problem is the high population density which has resulted in housing development in coastal areas [3].

2. MARINE TURTLES ALONG THE KERALA COAST

Marine Turtles have been reported from Kerala since the last many years. Mainly seven species of turtles are reported across the world, out of which five are found in India, and and four in Kerala, Olive ridley (*Lepidochelys olivacea*), Green turtle (*Chelonia mydas*), Leatherback (*Dermochelys coriacea*) and Hawksbill (*Eretmochelys imbricate*) [4-7]. But the most common among them along the Kerala Coast is the *Olive ridley* [8, 9].

Only one study has been conducted earlier in Kerala on Marine Turtles in September 2004. The study by Thanal Conservation Action and Information Network [10, 11]. was conducted by traversing the entire length of the Kerala Coast by the research team and interacting with fishermen, the coastal communities and general public to find out about turtles and their related details. The study identified the major beaches along the Kerala Coast from where turtles have been sited, the nesting beaches, local conservation efforts mainly in northern parts of the State, threats to turtles *etc*. The major threats to the turtles identified by the study include collection and sale of turtle eggs mainly along the northern parts of the State and capture, killing and consumption of turtle meat in the southern parts of the State.

3. IDENTIFICATION OF TURTLE NESTING BEACHES

The present study has tried to identify the natural turtle landing and nesting beaches in Kerala through interaction with the coastal people. In almost 90% of the locations, the people identified the species as *Olive ridley* except for an isolated sighting of Green Turtle from Thrissur. The major area covered through the survey which has some confirmed or unconfirmed reports of marine turtles includes:

- Trivandrum District: Pozhiyur, Poovar, Vizhinjam, Kovalam, Beemapalli - Poonthura, Veli - Shankhumukham, Perumathurai - Thazhampally, Anchuthengu and Edava - Odayam - Varkala beach.

- Kollam District: Kollam - Thangassery harbour, Kappil - Edava - Odayam and Mukkam – Pozhikkara.

- Alleppey District: Aleppey beach, Punnapra beach, Thottappally - Punnapra beach, Pallana beach, Thaikkal - Ottamasseri beach and Valiyazhikkal.

- Ernakulam District: Munambam and Cherai.

- Thrissur District: Azhikkode beach, Kothalam - Kaipamangalam beach, Nattika beach, Moothakunnam - Snehatheeram beach, Thalikulam - Nambikkadav beach, Vadanappilly beach, Blangad - Moonamkallu beach, Chavakkad - Puthenkadappuram beach, Edakkazhiyoor - Nalamkallu beach, Mannalamkunnu beach, Periyambalam beach.

- Malappuram District: Kadalundi (Anangadi), Muthiyam (Vallikkunnu), Parappanangadi - Alungal, Kettungal - Tanur, Unniyal, Koottayi, Pandayi - Nair Thodu, Koottayi Azhimukham, Ponnani - Marakkadavu, Puthuponnani - Munambam Jaram - Veliyancode - Palappetty (Perumbadappu).

- Kozhikode District: Chaliyam beach, Beypore - Calicut - Puthiyappa beach, Kappad beach, Thikkodi - Kolavipalam beach.

- Kannur District: Ezhimala - Ettikkulam beach, Choottad - Mattool beach, Meenkunnu - Chal beach, Payyambalam beach, Thottada - Ezharakadappuram beach, Muzhuppilangad beach.

- Kasargod District: Mavila - Thrikkaripoor beach, Thaikkadappuram - Azhithala beach, Ajanur - Hosdurg beach, Chittari beach, Pallikkara (Bakel) beach, Kottikulam beach, Odath - Kappil beach, Chembarikka beach, Kizhoor beach, Kasaba - Nellikunnu beach, Mogral - Kavugoly beach, Peruvad - Kumbala Beach, Shiriya - Ayila beach, Uppala - Benkkara beach, Manjeswaram beach, Thalappady - Kanwatheertha beach.

Except for Kolavipalam in Kozhikode and Neeleshwaram in Kasargod, there are no good records of the turtle numbers, eggs and hatchlings over the years. In these two areas, local conservation groups *i.e.* Theeram in Kolavipalam and Naithal in Neeleshwaram have been active over the past many years in turtle conservation and hence they have maintained good record of the turtles and their nesting details. A recent turtle protection initiative has been locally launched at

Chavakkad in Thrissur by a small NGO by the name of 'Green Habitat' under the leadership of a school teacher. They do not have much record of turtles and their nests, but they are trying to create awareness and sensitize the school children along the coastal schools towards turtle conservation.

The number of marine turtles coming to the beaches of Kerala has considerably reduced over the years due to a wide range of reasons - physical obstructions like rocky beaches, sea wall and tourism development. Further, habitat loss due to inappropriate afforestation activities by raising plantations and beach sand mining, killing of the turtles trapped in trawling nets, poaching for their meat, collection of eggs for human consumption, predation by dogs, jackals, mongoose *etc.* mainly by destroying the nests, increasing disturbance on the beaches by humans, artificial lighting *etc.* are also causes for the decline.

4. DIRECT AND INDIRECT THREATS

4.1. To Nesting Beaches (*e.g.* Beach Erosion, Casuarina Plantations, Sand Mining, *etc.*)

Casuarina plantations along the coast and sand mining are some of the major threats. Turtles are unable to dig nests on the sand due to the hindrance caused by roots of casuarinas. The demand for sand is high in Kerala and sand reserves in rivers are mostly exhausted. So there is a shift to beach sand mining and using the beach sand for mainline construction purposes after washing them with freshwater to get rid of the salinity. One such example is the beach in Kolavipalam which is fast shrinking due to this illegal activity.

4.2. To the Turtles

4.2.1. From Incidental Catch (Gill Nets and Trawl Nets)

Many of the turtles get caught in the trawl nets die because of lack of oxygen since they normally breathe by coming to the surface to take in air. Such turtles may also be killed by fishermen and are consumed as reported from Vizhinjam, Thankasseri – Kollam, Alleppey and Azheekkal [12, 13]. Turtles may also cause damage to nets leading to great losses to fishermen as reported from Calicut, Malappuram and some parts of Thrissur district also.

On the other hand, in Kottikulam area of Kasargod, the fishermen try to rescue the turtles caught in the trawl nets and release them back into the waters because of religious beliefs that turtles are considered as the reincarnation of Lord Vishnu.

4.2.2 Consumption for Meat and Eggs

As the turtles leave their flipper marks when they make their nests, lay eggs and go back to the sea, it is easy for the local people to find the nests and collect the eggs. The eggs are used for household consumption by the fishermen and the local community and also sold illegally. Nowadays the local conservation groups are coming up in many areas for the protection of the turtles and there is an increased awareness due to the stepping up of enforcement by the Kerala Forests and Wildlife Department personnel. But still fishermen confess that, collection of eggs is very much opportunistic.

Consumption of turtle meat is more common in the southern regions of Kerala especially in some areas where it is believed to be a cure for asthma [14].

The nature of selling is slightly different from area to area. In Vizhinjam, the turtles caught are kept alive by leaving them in the drain running alongside the harbour by securing one of the flippers by a rope to a pillar or heavy stone. Only when a buyer approaches them, they retrieve the turtle from the waterbody. Sometimes the turtles are put upside down in the harbour area and they are usually kept covered with dry banana leaves so that they are not seen easily.

Trawl operators and fish merchants are also very active in this trade.

From the coastal districts, sometimes the turtle meat goes to other districts like Kottayam, where it is in high demand. Predation of eggs by stray animals such as dogs is another major threat and even though only a few eggs may be consumed the entire nest is destroyed during the process.

4.2.3. From Impacts of Coastal Development (e.g. Seawalls, Ports, Fishing Harbours, Lighting, Tourism, etc.)

About 60-70% of coastline of Kerala has Sea walls,, which prevent the turtles from entering the beach for nesting, in spite of their long travelling from distant places.

Kerala has many fishing harbours and fish landing centres and such human activities disturb turtle nesting The noise and disturbances caused by mechanised boats also disturb nesting turtles. Many turtles get trapped in nets and die because the nets are not fitted with any Turtle Excluder Devices (TED).

5. DISCUSSION AND RECOMMENDATIONS (WHAT EFFORTS ARE NEEDED ON CONSERVATION OF MARINE TURTLES, WHO SHOULD BE INVOLVED)

Based on the various threats to marine turtles, the various recommendations include:

i. Consumption of meat and collection of eggs –Awareness programmes targeted at different communities right from grassroot level is necessary. Also enforcement should be strengthened by capacity building of the Forest Department personnel on such subjects, more vigilance through patrolling and raids wherever necessary, initiation and strengthening of local informants *etc.*

ii. Mortality in trawling nets – Turtle Excluder Device (TEDs) should be installed in all trawling nets and proper compensation should be given to fishermen suffering financial losses due to damages of fishing nets.

iii. Predation – Patrolling of beaches should be intensified to avoid predation by stray animals.

iv. Loss of habitat – Coastal afforestation should be done in such a manner that the movement of turtles are not hindered.

v. Coastal development activities Construction of sea walls, implementation of tourism developmental projects and illumination in beaches should be done with minimum disturbance to nesting sites.

vi. Polluting activities like waste dumping, effluent disposal from industries *etc.* should be strictly regulated.

vii. Coastal Eco Development Committees (EDCs) should be initiated so that the local people develop a sense of awareness.

viii. The Kerala Forests and Wildlife Department should have a separate division for wildlife outside forests including marine turtles.

ix. Proper coordination should be ensured between the various enforcement agencies including the Kerala Forests and Wildlife Department, Police, Coast Guard, Marine Enforcement, Mining and Geology *etc.*, with the State Biodiversity Board.

ACKNOWLEDGEMENTS

We would like to thank all the people along the Kerala Coast with whom we had some level of formal or informal interaction – fishermen, trawl operators, coastal communities, youth clubs, representatives of fishermen/trawl operators societies, representatives of local self government departments, enforcement personnel *etc.* We would also like to place on record our heartfelt thanks and gratitude to Salam Usman, Mohammed Shafeeq, Arunjith P.S., Nature Club Members from MES College, Ponnani, Arif P., Mohammed Faisal, Nidhish, Nature Club Members from PSMO College, Thirurangadi and Lekshmi Nambiath, Amrita College of Engineering, Kollam who were volunteers in the survey along the coast. A special word of thanks goes to Mr. Maqbool T.K., Lecturer, Department of Zoology, MES College, Ponnani for taking the lead in arranging the volunteers.

CONFLICT OF INTEREST

The authors confirm that this chapter contents have no conflict of interest.

REFERENCES

[1] ACCESS Knowledge Series. Catalysing Markets through Collectives: Experiences from the Allied Sector. Sitaram Rao Livelihoods India Case Study Competition 2011; pp. 46.

[2] KSCSTE. Muthunayagam AE, Ed. Environmental Impact of Tsunami in the Kerala Coast Government of Kerala ENVIS Centre Publication 2006; pp. 6-8.

[3] Sreekala SE, Baba M, Muralikrishna M. Shoreline changes of Kerala coast using IRS data and aerial photographs. Indian Journal of Marine Science 1998; 27: 144-148.

[4] BOBLMEP. National Report of Sri Lanka on the Formulation of a Transboundary Diagnostic Analysis and Strategic Action Plan for the Bay of Bengal Large Marine Ecosystem Programme 2003; pp. 27.

[5] Cameron TH. Notes on turtles. Journal of the Bombay Natural History Society 1923; 29: 299-300.

[6] Kar CS, Bhaskar S. Status of sea turtles in the eastern Indian ocean. In: Bjorndal K, Ed. Biology and conservation of sea turtles. Smithsonian Institution Press: Washington D. C. 1982; pp. 365-372.

[7] Smith MA. The fauna of British India including Ceylon and Burma. Reptilia and Amphibia. Vol 1. Loricata, Testudines. Taylor and Francis Ltd. London 1931; pp 185.

[8] Bhaskar S. The distribution and status of sea turtles in India. In: Silas EG, Ed. Proceedings of the Workshop on Sea turtle Conservation. Central Marine Fisheries Research Institute, Special Publication 1984; 18: 21-35.

[9] Bhupathy S. Monitoring of marine turtles along the Kerala and Tamil Nadu coasts. Indian Ocean Turtle Newsletter 2007; 5: 1-9.

[10] Dileepkumar N, Jayakumar C. Field study and networking for the turtle conservation in Kerala - A GOI-UNDP sea turtle project report 2002.

[11] Dileepkumar N, Jayakumar C. Sea turtles of Kerala. In: Shanker K, Choudhury BC, Eds. Marine turtles of the Indian subcontinent. Universities Press (India) Private Ltd. 2006; pp. 137-140.

[12] Krishna Pillai S, Suersh KK, Kannan P. Leatherback turtle released into the sea at Kovalam, Kerala, India, Kachhapa 2003; 9: 5-6.

[13] Krishna Pillai S. Accidental catch of green turtle Chelonia mydas in shore seine at Kovalam near Vizhinjam, South Kerala. Fishing Chimes 2004; 24(3): 57.

[14] Shanmugasundraram P. Turtle industry. Indian Seafoods 1968; 6: 18-19.

CHAPTER 19

Synchronous and Rhythmic Light Display by a Panoramic Congregation of Fireflies at Varagaliar, Anamalai Tiger Reserve

Mathish Nambiar-Veetil*, Selvaraj P., Blessan Santhosh George, Ganesan M., Raghunath T. P. and Krishnakumar N.

Institute of Forest Genetics and Tree Breeding, Forest Campus, R. S. Puram, Coimbatore, India

Abstract: Synchronous flashing of a large congregation of fireflies observed at Varagaliar, located in the Anamalai Tiger Reserve in the Tamil Nadu part of Western Ghats (10°25.09' N, 76°51.94' E, Elevation 2100 ft) is reported here. Typically, the fireflies on the topmost part of the tree crown flash a yellowish green light, almost immediately followed by similar light flashes serially by the fireflies below. The flashing pattern in each tree repeated itself cyclically, and other neighbouring trees had similar but unique synchronous displays. After a few cycles, fireflies in all the trees flash three times in unison as if appreciating their spectacular performance. The whole forest landscape resonated with pulsating waves of multiple kaleidoscopic flashes of light from several clusters of fireflies until dawn. This annual spectacle is observed only during the month of April, before the onset of the south-west monsoon, perhaps coinciding with the mating season of the adult insect that lives for about a month. Synchronous fireflies have also been reported elsewhere from Southeast Asia, the Great Smoky Mountains National Park, Elkmont, USA, and Brazil. Conserving this unique phenomenon at Varagaliar for posterity, necessitates a better understanding of the species composition of the fireflies involved in this phenomenon, the reasons for their synchronous behaviour, the geographical area that is congregated by these insects, the unique plant species composition, and the ecology and ecoclimate of this habitat that makes it an exclusive niche for congregation of synchronous fireflies.

Keywords: Anamalai, annual spectacle, conservation, cyclic, ecotourism, glow worms, mating, monsoon, synchronous fireflies, vegetation, Western Ghats.

1. INTRODUCTION

The remnant forests of the Western Ghats provide a safe sanctuary for unique and amazing life forms that were once very common elsewhere. The ability of fireflies to bioluminesce has always fascinated mankind, and the understanding of the

***Corresponding author Mathish Nambiar-Veetil:** Institute of Forest Genetics and Tree Breeding, Forest Campus, R. S. Puram, Coimbatore, India; Tel: 91 422 2484125; E-mail: mathish@icfre.org

Laladhas K.P., Oommen V. Oommen and Sudhakaran P.R. (Eds.)

molecular basis of this remarkable phenomenon has value in biotechnology. A number of anecdotal reports point to the decreasing population of fireflies in human habitations. This observational paper describes a chance sighting of the spectacular phenomenon of synchronous flashing by a panoramic congregation of fireflies (Lampyridae, Coleoptera) at Varagaliar, Anamalai Tiger Reserve, in Tamil Nadu, India.

2. CYCLIC PATTERN OF FIREFLIES AT VARAGALIAR

Varagaliar is approachable from Topslip near Pollachi in Tamil Nadu, and also from the Sholayar power house in Valparai, Tamil Nadu, and is located 5 km away from the Parambikulam reservoir near the Tamil Nadu- Kerala border. The two days visit to Varagaliar had been envisaged for assessment of vegetation as a part of the *Indian Council of Forestry Research and Education* (ICFRE) study on "Reassigning forest type of India for better management of forests in India". The sighting of the synchronous flashing by a large congregation of fireflies was observed in the late evening hours in front of the forest rest house at Varagaliar (10°25.09' N, 76°51.94' E, elevation 2100 ft).

It was around 6.30 pm, on 21st April 2012, in the immediate vicinity of the valley in front of the solar power -lit forest rest house at Varagaliar (Fig. **1**), when numerous fireflies (Lampyridae, Coleoptera) were observed flying around. Soon the innumerable fireflies had congregated on the trees, and then it was noticed that the flickering light emitted by the stationary fireflies had a cyclic and a synchronous pattern. These synchronous flash patterns had characteristic features. The firefly on the topmost part of the tree crown flashes a yellowish green light, almost immediately followed by similar light flashes serially by the fireflies below. This orchestrated flashing pattern repeated itself cyclically, and other neighbouring trees had similar but unique synchronous displays. The whole visible forest landscape resonated with pulsating waves of multiple kaleidoscopic light flashes from several clusters of fireflies. After a few cycles, fireflies in all the trees characteristically flashed three times in unison. Until before dawn, the whole forest seemed to be celebrating in anticipation of the advent of the southwest monsoon downpour in May, and the panoramic spectacle rivalled the splendour of flickering stars in the sky. The forest watcher, who stays very near to the forest guest house, observed that this annual spectacle was noticed only during the month of April, perhaps coinciding with the mating season of the adult insects that lives for about a month. Varagaliar is located in the transition zone of the Teak plantations and the pristine natural forests of the Anamalai tiger reserve. Few of the common trees found in and around the rest house included *Tectona grandis* plantations, *Bambusa bambos, Terminalia crenulata, Dalbergia latifolia, Lagerstroemia speciosa, Trewia polycarpa, Cassia fistula, Swietenia mahagoni, Hydnocarpus pentandra, Alstonia scholaris, Cycas circinalis, Macaranga peltata* and *Calophyllum* sp. The birds observed include *Ocyceros griseus, Gracula religiosa, Vanellus indicus, Pycnonotus* spp. and *Acridotheres fuscus.* Less than half a km from the rest house is a small stream frequented by Gaur and Sambar deer herds.

Figure 1: The forest landscape canvas in front of the Varagaliar Forest Rest house.

This phenomenon is also known to occur near the Parambikulam reservoir in Kerala (report based on personal communications with forest staff), which is just 5 km away from the forest guest house at Varagaliar. It would however, be pertinent to record that synchronous flashing was not noticed the previous night in the small number of fireflies found near the Sholayar power house (10°21.45' N, 76°52.679' E, elevation 1963 ft), which is about 7.5 km from Varagaliar. Synchronous fireflies have also been reported elsewhere from Southeast Asia [1], the Great Smoky Mountains National Park, Elkmont, USA [2], and Brazil [3]. The common Southeast Asian species of synchronous fireflies include *Pteroptyx malaccae, P. tener* and *P. valida,* and in North America they include *Photinus carolinus, P. pyralis, P. consimilis,* and *P. jamaicensi* [4]. Along the mangrove forests in the estuary of the Selangor River at Kuala Selangor, Malaysia, the nectaries of the young trees of *Sonneratia caseolaris*, provide the required nutrition for adult *P. tener* [5].

The Morse-code like species specific synchronous flashes have been associated with the behaviour of a cluster of male fireflies courting the same female [4], a plausible reason cited for synchronous fireflies to migrate away from human habitations polluted with more powerful artificial lights. Synchronous behaviours are common in cicadas, crickets and frogs, which use synchronized sound to communicate amongst the individuals. Mathematical models like the Kuramoto model have been used to capture such biological synchronous behaviour in congregating individuals [6]. Elsewhere in the world, synchronous light display of fireflies draws numerous visitors to the Great Smoky Mountains National Park, Elkmont, USA, and to the Kuala Selangor National Park, Malaysia. In Malaysia, the local village cooperatives operate boat service for the visitors to view the spectacle. Making Varagaliar as an excursion spot for the public to experience this biological marvel would rather be a contentious issue, considering the fact that such phenomenon happens only in few parts of the world, in addition to anecdotal reports of decreasing firefly populations near human habitations.

CONCLUSION

Conserving this phenomenon at Varagaliar for posterity, necessitates a better understanding of the species composition of the fireflies involved in this phenomenon, the reasons for their synchronous behaviour, the geographical area that is congregated by these insects, the unique floral and faunal composition, and the ecology and ecoclimate of Varagaliar that makes it an exclusive niche for congregation of synchronous fireflies. Any decision for the present generation to view the spectacle should be made only after taking into consideration the findings from such a detailed study, and after provision of appropriate minimalistic, non- polluting (light/sound/air/water) infrastructure and arrangements that do not make the unique fauna of Varagaliar to migrate elsewhere.

ACKNOWLEDGEMENTS

We would like to thank the Indian Council of Forestry Research and Education, Dehra Dun for providing funding support under the study on "Reassigning forest type of India for better management of forests in India". We are also grateful to the Tamil Nadu Forest Department, for the permission and logistical support for visiting Varagaliar.

CONFLICT OF INTEREST

The authors confirm that this chapter contents have no conflict of interest.

REFERENCES

[1] Smith HM, Synchronous flashing of fireflies. Science, 1935; 82: 151–152.
[2] Copeland J, Moiseff A. The occurrence of synchrony in the North American firefly *Photinus carolinus* (Coleoptera: Lampyridae). J. Insect Behav, 1995; 8: 381–394.
[3] Viviani VR. Fireflies (Coleoptera: Lampyridae) from southeastern Brazil: habitats, life history, and bioluminescence. Ann Entomol Soc Am. 2001; 94: 129-145.
[4] Copeland J, Moiseff A, Faust L. Landing distance in a synchronic North American firefly. Physiol Entomol. 2008; 33: 110–115.
[5] Nallakumar K. The synchronously flashing aggregative fireflies of peninsular Malaysia. Biodiversity, 2003; 4: 11-16.
[6] Acebrón JA, Bonilla LL, Vicente CJP, Ritor F, Spigler R. The Kuramoto model: A simple paradigm for synchronization phenomena. Phys. Rev., 1998; E 57, 5287.

Chapter 20: Identification, Purification and Mass Spectrometric Characterization of Two Novel Cyclic Peptides from Clitoria ternatea

CHAPTER 20

Identification, Purification and Mass Spectrometric Characterization of Two Novel Cyclic Peptides from *Clitoria ternatea*

Sudarslal S.[1,2,*], Balaram P.[1] and Laladhas K. P.[1,3,4,*]

[1]Molecular Biophysics Unit, Indian Institute of Science, Bangalore, Karnataka, India; [2]Amrita School of Biotechnology, Amrita Vishwa Vidyapeetham, Kollam, Kerala, India; [3]Biomolecular Research Unit, Department of Zoology,St. Stephen's College, Pathanapuram, Kollam, Kerala, India and [4]Kerala State Biodiversity Board, L-14, Jai Nagar, Medical College P.O., Thiruvananthapuram Pin- 695 011, Kerala, India

Abstract: Kerala has a rich heritage of plant diversity and health traditions. The history of drug discovery reveals that the ethno-botanical approach is one of the most productive methods to screen bioactive compounds. Many drugs currently in use were developed after scientists began to analyze the chemical constituents of plants. Exploration of the medicinal uses of plants from traditional knowledge coupled to scientific research methodologies may be able to determine whether such plants provide therapeutically relevant drug candidates or not. *Clitoria ternatea*, a plant from the family Fabaceae is a slender, perennial climber used as an anti-inflammatory agent in folklore medicine. In an effort to identify biologically active molecules from *Clitoria ternatea*, we have isolated two novel, disulfide rich cyclic peptides using reversed phase high performance liquid chromatography (RP HPLC). Masses of these peptides were measured as 3058.8 Da and 3074.4 Da by an electrospray ionization mass spectrometer (ESI MS). Partial sequences of the peptides were assigned through manual *de novo* sequencing approach using tandem mass spectrometric data obtained from a matrix assisted laser desorption ionization time of flight mass spectrometer (MALDI-TOF MS).

Keywords: Clitoria ternatea, Cyclic Peptide, de novo sequencing, Electrospray, MALDI, Carboxamidomethylation.

1. INTRODUCTION

Cyclic peptides belong to a family of biologically active plant peptides and are composed of approximately 30 amino acid residues with head-to-tail cyclized

***Corresponding authors Sudarslal S. and Laladhas K.P.:** Molecular Biophysics Unit, Indian Institute of Science, Bangalore, Karnataka, India; Tel: 0471 - 2559134 (off.); Fax: 0471 – 2448234; Mob: 9447271034; E-mails: sudarslal@am.amrita.edu, laladhas@gmail.com

Laladhas K.P., Oommen V. Oommen and Sudhakaran P.R. (Eds.)

peptide backbone and three interchain disulfide bonds [1, 2]. This distinctive structural feature known as cyclic cysteine not only confers them exceptional chemical, thermal and enzymatic stability but also makes them attractive scaffolds for drug design and delivery [3]. These peptides, considered to play an important role in plant defense mechanism, exhibit a wide range of biological activities such as uterotonic, antimicrobial, insecticidal and anti-HIV properties [4, 5]. Though a large number of cyclic peptides were reported from plants belonging to Violaceae, Rubiaceae and Cucurbitaceae families [6], only few reports were published from families such as Apocynaceae and Fabaceae [7, 8]. Here, we report combined use of high performance liquid chromatography (HPLC) and MALDI MS to identify, purify and structurally characterize two novel cyclic peptides from *Clitoria ternatea*, a perennial plant that belongs to the Fabaceae family.

2. PEPTIDE EXTRACTION

Shoot regions of *Clitoria ternatea* were collected, cleaned well and shade dried. The dried materials were kept in an incubator at 37 °C and powdered using a motorized mixer. The powdered sample was subjected to methanol extraction using a Soxhlet apparatus, followed by steeping with petroleum ether to remove chlorophyll. The methanol fraction was concentrated on a rotary evaporator and subjected to HPLC purification. The plant extract was preliminarily purified on a HP 1100 series HPLC system, using a C-18 reversed phase column (Jupiter, Phenomenex, 10 × 250 mm, 4 μ particle size, 90 Å pore size). Further purification was effected on a C-18 reversed-phase column (Zorbax, 4.6 × 250 mm, 5 μ particle size, 300 Å pore size). A linear gradient was employed with distilled water and acetonitrile containing 0.1% trifluoroacetic acid as mobile phase. UV absorbance was monitored at 226 nm.

2.1. Chemical Modifications

2.1.1 Reduction and Alkylation

The purified peptide fractions were dissolved in 0.1 M NH_4HCO_3 buffer, pH 8.0. For the reduction, 200 mM dithiothreitol stock solution was added to get a final concentration of 8 mM and incubated at 37 °C for 2 hours. To this solution, appropriate iodoacetamide stock solution was added to get a final concentration of 20 mM. The resultant mixture was incubated at room temperature in the dark for 45 minutes followed by desalting using a C-18 reversed phase column.

2.1.2. Trypsin Digestion

The purified, carboxamidomethylated peptide fractions were suspended in 0.1 M NH_4HCO_3 buffer, pH 8.0 and digested with TPCK-treated trypsin (Sigma, USA) for three hours at 37 °C. The digest was directly analyzed by MALDI MS.

2.1.3. V8 Protease Digestion

The purified, carboxamidomethylated and trypsin digested peptide fractions suspended in 0.1 M NH_4HCO_3 buffer, pH 8.0 were mixed with V8 protease and incubated at 37 °C for overnight.

2.1.4. Endoproteinase Arg-C Digestion

The purified, carboxamidomethylated peptide fractions dissolved in 0.1 M NH_4HCO_3 buffer, pH 8.0 were mixed with Endoproteinase Arg-C and incubated at 37 °C for overnight.

2.1.5. Mass Spectrometry

MALDI mass spectra were acquired on an Ultraflex TOF/TOF system (Bruker Daltonics) fitted with a nitrogen laser of wavelength 337 nm. The peptide solutions were mixed with equal amount of matrix solution (α-cyano 4-hydroxy cinnamic acid) and dried at room temperature. Spectra were recorded in reflectron, positive ionization mode. The MS/MS spectra (tandem mass spectra) were recorded employing the LIFT technology. Electrospray ionization mass spectra were recorded using a Hewlett Packard 1100 MSD series mass spectrometer fitted with a single quadrupole mass analyzer.

3. CYCLIC PEPTIDES FROM *CLITORIA TERNATEA*

Two cyclic peptides, named CTMA and CTMB were isolated and purified from *Clitoria ternatea*. Fig. **1**. shows the HPLC profile of the *Clitoria ternatea* shoot extract. Precise molecular masses of CTMA and CTMB were determined by ESI MS and are found to be 3074.4 Da and 3058.8 Da, respectively, where the two peptides differed by a mass unit of 16 Da. Reduction with dithiothreitol and alkylation with iodoacetamide (carboxamidomethylation) of CTMA and CTMB yielded 348 Da more to the intact molecular masses. The increases in masses were a clear indication of the presence of 3 disulfide bridges in the peptides (carboxamidomethylation yields an increase in mass of 116 Da/disulfide bond). Since there were many reports on plant-derived cyclic peptides (N to C cyclized peptides), we have carried out various enzymatic digestion protocols to see whether the peptides isolated from *Clitoria ternatea* were cyclic or not. Mass spectrometric analysis of the carboxamidomethylated peptides after trypsin digestion showed an increase in mass of 18 Da, clearly indicating that CTMA and CTMB are cyclic in nature and harbouring one trypsin cleavage site (trypsin cleaves peptides at the C-termini of arginine and lysine amino acid residues) (Fig. **2**). However treating the peptides with endoproteinase Arg-C (this enzyme cleaves C-termini to arginine residues) didn't yield any mass shift indicating that the peptides contain no arginine residues.

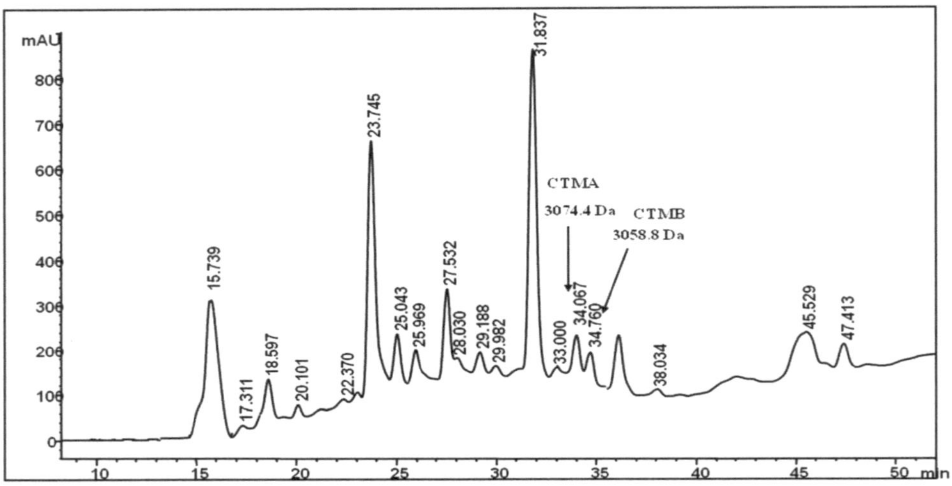

Figure 1: HPLC profile of crude shoot extract of *Clitora ternatea*. The peaks labeled as CTMA and CTMB were characterized. The molecular masses detected by ESI MS are indicated above the respective chromatographic peaks.

Both the cyclic peptides were carboxamidomethylated and trypsin digested to derive primary amino acid sequences. Thus linearized peptides were HPLC purified and subjected to various proteolytic cleavages followed by tandem mass spectrometric data collection. V8 protease digestion of CTMA and CTMB respectively yielded two prominent fragment masses at 2612.4 Da and 2596.7 Da. It is clear from this observation that the mass difference of 16 Da between the native CTMA and CTMB is retained between the V8 protease generated peptide fragments too. By following Biemann's nomenclature [9], tentative sequences of CTMA and CTMB peptide fragments were derived through manual *de novo* sequencing approach. Both the tandem mass spectra, by and large yielded identical sequences as CJPDASCSCDPVYCTGNTDE, where J can be either L or I (Fig. **3**) shows the observed fragment ions along with the assignments of b-ion series). On a closer inspection, it can be seen that the b_2 ions of the two tandem mass spectra carry the same mass difference of 16 Da that was originally observed between native CTMA and CTMB. But, due to the poor fragment ion quality, no conclusive data could be generated for assigning the missing N-termini amino acid sequences. However, by considering various amino acid combinations, it can be inferred that the b_2 ion at 293.1 Da (the ion obtained from the tandem mass spectrum of CTMA after V8 protease digestion) could correspond to the dipeptide, EY. Similarly, the ion at 276.7 Da (the b_2 ion obtained from the tandem mass spectrum of CTMB after V8 protease digestion) represents three possible dipeptide combinations, *viz*. IY, LY or EF.

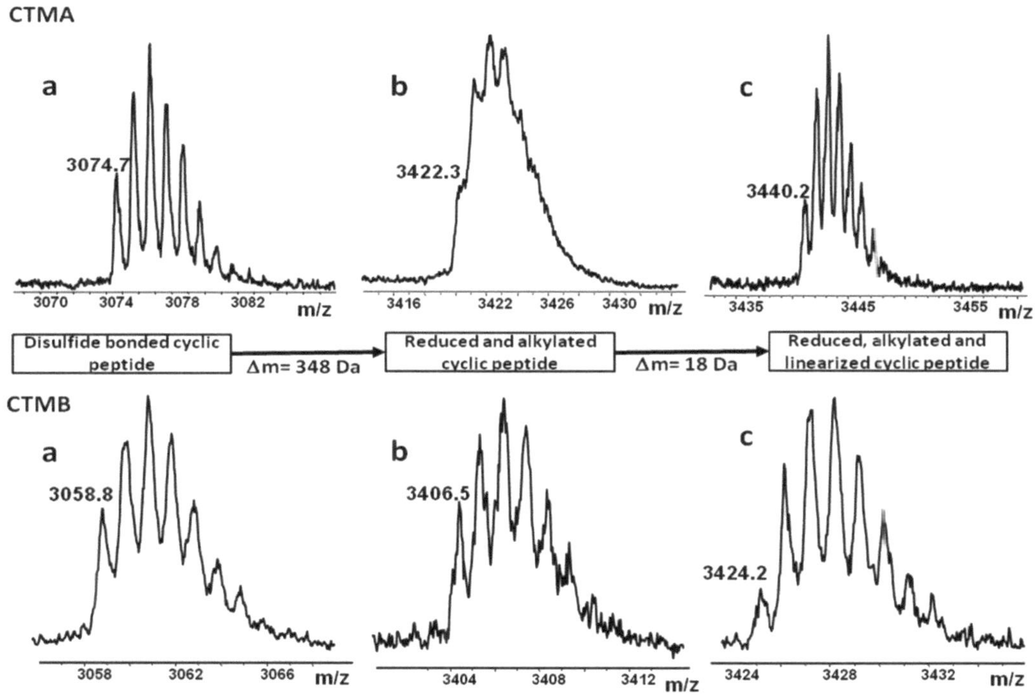

Figure 2: MALDI mass spectra showing masses of CTMA and CTMB in the native form, (a); after carboxamidomethylation, (b) and carboxamidomethylated peptides after trypsin digestion, (c). The respective masses are also labeled.

On the other hand, the tandem mass spectra of carboxamidomethylated and trypsin digested CTMA and CTMB, with the help of V8 protease fragmentation data, yielded identical sequence as CJPDASCSCDPVYCTGNTDECSGVCHK (Fig. **4**), which also failed to provide comprehensive information on the N-termini

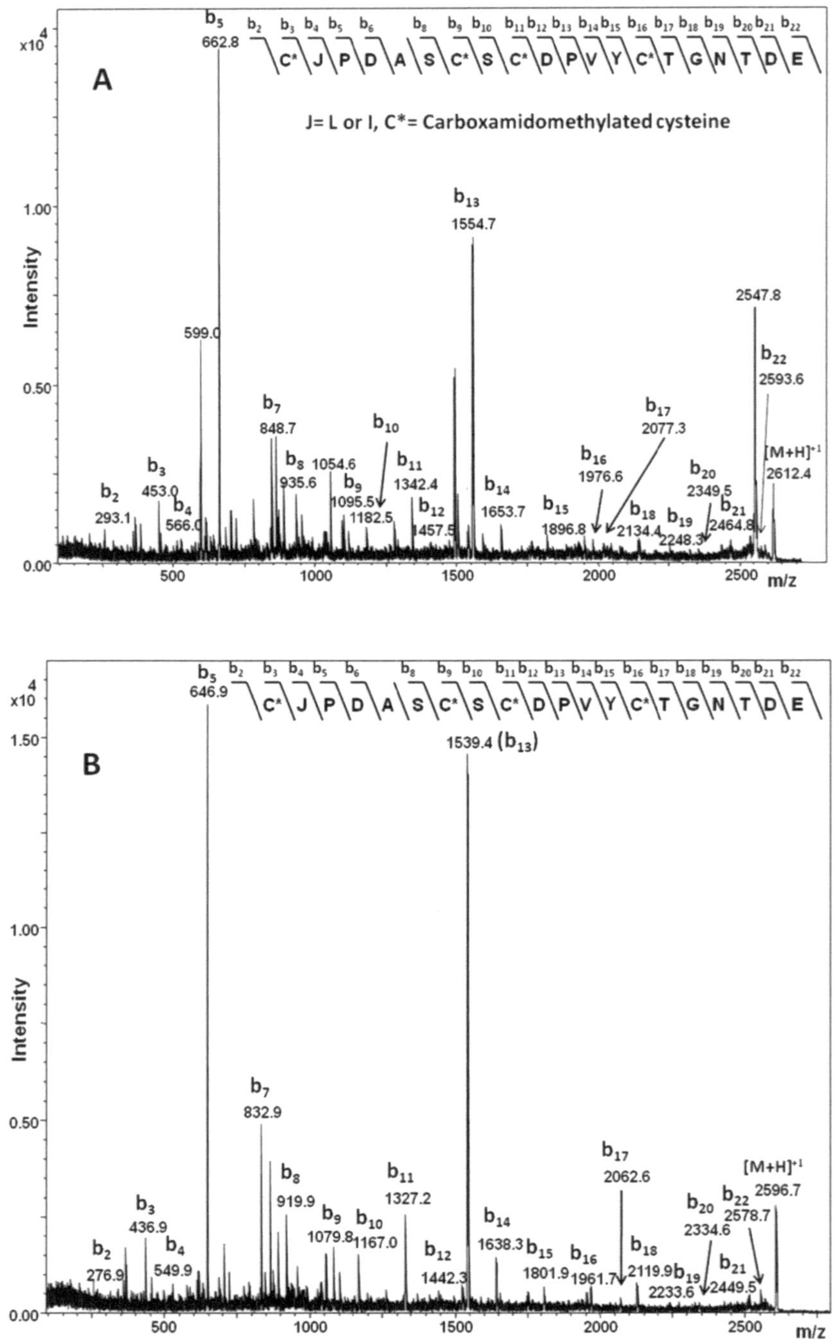

Figure 3: MALDI MS/MS of CTMA (A) and CTMB (B) after carboxamidomethylation, trypsin digestion and V8 proteolytic cleavage. Monoisotopic m/z values of the observed b-ions are marked. The manually assigned sequences through *de novo* approach are also included as insets.

amino acids of the linearized peptides. With this ambiguity in place, the final determined sequences of CTMA and CTMB are XXCJPDASCSCDPVYCTGNT DECSGVCHK, where XX can be EY for CTMA and IY, LY or EF for CTMB.

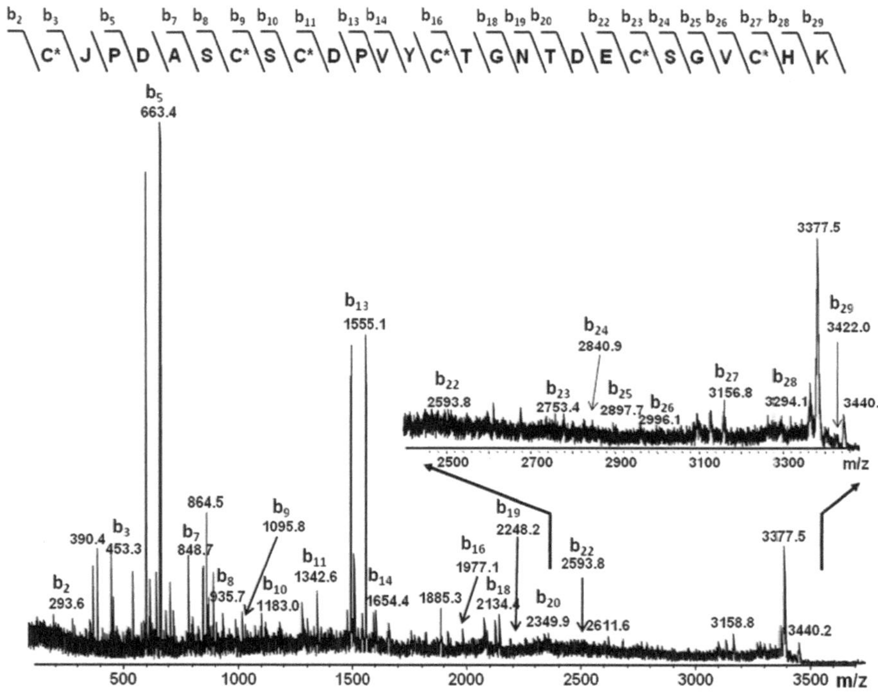

Figure 4: MALDI MS/MS of CTMA after carboxamidomethylation and trypsin digestion. The inset shows highlighted region of the tandem mass spectrum and the manually assigned sequence.

In this study, we describe the identification, purification and mass spectrometry based characterization of two novel cyclic peptides from the shoot extract of *Clitoria ternatea*. The primary amino acid sequences of the peptides were derived through tandem mass spectrometry data based manual *de novo* sequencing approach. The sequences of the peptides were confirmed through proteolytic digestion protocols.

ACKNOWLEDGEMENTS

The mass spectrometry facility at IISc., Bangalore supported under the Proteomics Program funded by the Department of Biotechnology, Government of India is gratefully acknowledged for providing the mass spectrometry data.

CONFLICT OF INTEREST

The authors confirm that this chapter contents have no conflict of interest.

REFERENCES

[1] Colgrave ML, Poth AG, Kaas Q, Craik DJ. A New ''Era'' for Cyclotide Sequencing. Biopolymers 2010; 94(5): 592-601.

[2] Sudarslal S, Romanuka J, Billeter M, Skjeldal L, Emmett MR., Nilsson CL, Marshall AG. Structural characterization of an unusually stable cyclic peptide, kalata B2 from *Oldenlandia affinis*. Biochimica et Biophysica Acta 2006; 1764: 1568-1576.

[3] Ireland DC, Colgrave M, Craik DJ. A novel suite of cyclotides from *Viola odorata*: sequence variation and the implications for structure, function and stability. Biochemical Journal 2006; 400: 1-12.

[4] Pränting M, Lööv C, Burman R, Göransson U, Andersson DI. The cyclotide cycloviolacin O2 from *Viola odorata* has potent bactericidal activity against Gram-negative bacteria. Journal of Antimicrobial Chemotherapy 2010; 65: 1964-1971.

[5] Mylne JS, Wang CK, van der Weerden NL, Craik DJ. Cyclotides are a Component of the Innate Defense of *Oldenlandia affinis*. Biopolymers 2010; 94(5): 635-646.

[6] Gould A, Ji Y, Aboye TL, Camarero JA. Cyclotides, a novel ultrastable polypeptide scaffold for drug discovery. Current Pharmaceutical Design 2011; 17(38): 4294-4307.

[7] Gruber CW, Elliott AG, Ireland DC, Delprete PG, Dessein S, Göransson U, Trabi M, Wang CK, Kinghorn AB, Robbrecht E, Craik DJ. Distribution and evolution of circular miniproteins in flowering plants. Plant Cell 2008; 20(9): 2471-2483.

[8] Poth AG, Colgrave ML, Philip R, Kerenga B, Daly NL, Anderson MA, Craik DJ. Discovery of cyclotides in the fabaceae plant family provides new insights into the cyclization, evolution, and distribution of circular proteins. ACS Chemical Biology 2011; 6(4): 345-355.

[9] Biemann K. Sequencing of peptides by tandem mass spectrometry and high-energy collision-induced dissociation. Methods in Enzymology (Eds) 1990; 193: 455-479.

CHAPTER 21

Satellite Based Spectral Signatures of Coastal Wetlands and Mangrove Vegetation of Kasaragod, India

Mohammed-Aslam M. A.*

Department of Geology, Central University of Karnataka, Gulbarga, Karnataka, India

Abstract: The coastal wetlands form an important component of coastal ecosystem and is subjected to human interventions. Majority of the wetlands located at the coastal tracts are characterized by dense mangrove covers. Due to the complex environment and fragile nature of these systems, they are listed among the most threatened resources. Increased awareness and planned plantation activities have shown good tendencies with the objectives of conservations of mangroves in Kasaragod. Application of satellite remote sensing techniques is found to be very effective in delineating the mangroves systems, which provides vital information for best management practices. Satellite data has been employed in the present investigation to characterize the land cover features and retrieving the spectral responses of wetland vegetations of coastal regions of Kasaragod. The outcome of study suggests the best methodology for mangrove vegetation analysis and provides systematic approach for the conservation and management of the wetland vegetation of Kasaragod.

Keywords: Conservation, Kasaragod, management, remote sensing, salt tolerant ecosystem, wetlands.

1. INTRODUCTION

Mangroves represent the highly fragile salt tolerant coastal ecosystem located at the tropical and subtropical regions. They undergo continuous changes due to several environmental stresses. Remote sensing provides an effective means for delineating and mapping of wetlands and mangroves at a faster mode [1-9]. Increased awareness and planned afforestation activities have improved the conservation of mangroves in Kasaragod. Remote sensing analysis has been adopted in this study to map the mangroves of Pallom backwater of Kasaragod.

The present study has highlighted the importance of remotely sensed data for

***Corresponding author Mohammed-Aslam M. A.:** Department of Geology, Central University of Karnataka, Gulbarga, Karnataka, India; Tel: +91-9482338112; E-mail: maslam.in@gmail.com

Laladhas K.P., Oommen V. Oommen and Sudhakaran P.R. (Eds.)

inventorying, mapping and monitoring of mangroves of Kasaragod. Kasaragod district is endowed with relatively good areas occupied by wetlands (Table **1**).

Table 1: Wetland area in Kasaragod district [10]

Wetland category	No of wetlands	Total wetland area	% of wetland area	Area in ha Post monsoon	Area in Ha pre monsoon
Waterlogged	25	155	2.05	57	30
River stream	55	5007	66.22	4751	4244
Tanks/Ponds	8	35	0.46	1	1
Lagoons	1	1775	23.48	1632	1626
Sand/Beach	14	443	5.86	0	0

Detailed study has been conducted at Pallom backwater of Kasaragod coastal area. High resolution images of Google Earth have been used in the present study to delineate the areas occupied by mangroves. Images archived on 22nd November 2004 and 30th November 2009 have been used for the present study. The areas occupied by mangroves have been visually interpreted to obtain the accurate areal extent of the mangroves (Fig. **1**). ArcGIS has been used for onscreen digitization and subsequent visual interpretation of images. It has been found that significant changes have been noticed within a span of 5 years of time.

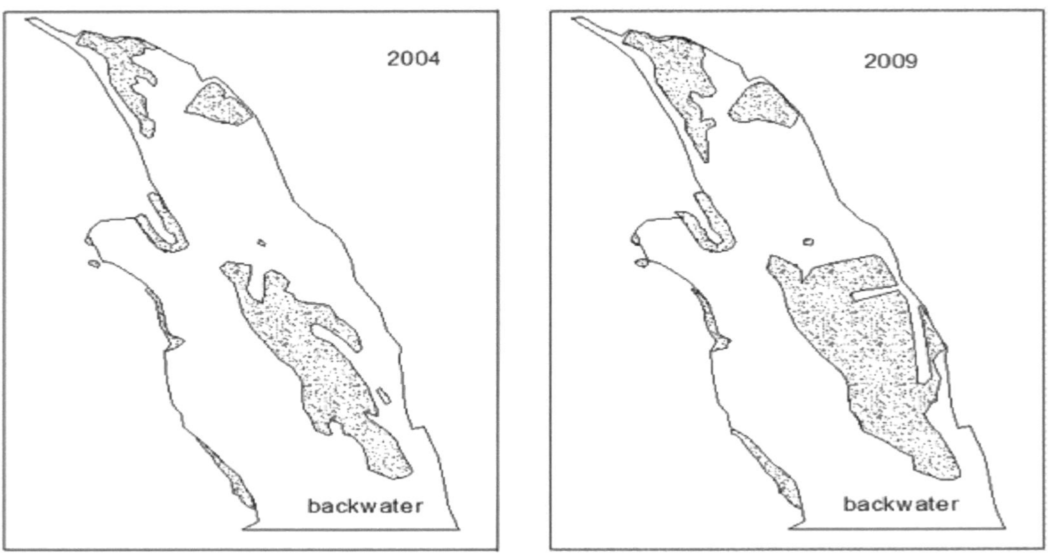

Figure 1: Area occupied by mangrove vegetation in Kasaragod in the years 2004 and 2009. The hatched area indicate the mangrove vegetation.

2. MANGROVE VEGETATION IN KASARGOD

Majority of the wetlands located at the coastal tracts are characterized by dense mangrove covers. Due to the complex environment and fragile nature of these systems, they are listed among the most threatened resources. Increased awareness and planned plantation activities with the objectives of conservation of mangroves has shown good tendencies. Satellite data have been employed in the present investigation to characterize the land cover features and for retrieving the spectral responses of wetland vegetations of coastal regions of Kasaragod.

Remote sensing method is highly useful to detect mangrove vegetation. Multi-temporal GoogleEarth images provide an easy way of conducting studies using visual image analysis. Onscreen visual interpretation of 2004 and 2009 done to detect the mangrove changes at Pallom backwater of Kasragod, showed significant increase in 5 year's time. The main reason for the increased mangrove cover is due to the government and community participation and movements for afforestation in this area. The outcome of study suggests the best methodology for mangrove vegetation analysis and provides systematic approach for the conservation and management of the wetland vegetation of Kasaragod.

ACKNOWLEDGEMENTS

Declared none.

CONFLICT OF INTEREST

The author confirms that this chapter contents have no conflict of interest.

REFERENCES

[1] Badhwar GD. Automatic corn soybean classification using landsat mss data. ii. early season crop proportion estimation. Remote Sensing of Environment 1984; 14(1-3): 31-37.

[2] Green EP, Mumby PJ, Edwards AJ, Clarke CO. A review of remote sensing for the assessment and management of tropical coastal resources, International Journal of Marine Environment 1996; 24: 1-40.

[3] Imbrernon J. Pattern and development of landuse changes in Kenyan highlands since 1950s. Agri. Ecosyst. Environ. 1999; 76: 67-73.

[4] Nayak S, Bahuguna A. Application of remote sensing data to monitor mangroves and other coastal vegetation of India. International Journal of Remote Sensing 2001; 30(4): 195-213.

[5] Mohammed-Aslam MA. Thrust Areas of Geospatial Applications in Fragile Ecosystems and a Case Study. Special Issue of Eco-chronicle 2010; pp. 74-78.

[6] Mohammed-Aslam MA, Abdussalam AK. GIS based Assessment for the need of conserving Biodiversity of south western low lands of India in relation to the terrain characteristics. J Mill. Zoo. 2011; 12: 72-73.

[7] Mohammed-Aslam MA, Gopinathan V, Sreemathi Gopinath K, Abdussalam AK, Madhusudhanan K. Interdisciplinary Approach in the Geospatial Applications of Fragile Ecosystems. Special Issue of Eco-chronicle 2010; pp. 62-66.

[8] Manish Sharma, Rashmi Gupta, Deepak Kumar and Rajiv Kapoor. Efficacious approach for satellite image classification. Journal of Electrical and Electronics Engineering Research 2011; 3(8):143-150.

[9] Vidyasagaran. K, Ranjan. M.V, Maneeshkumar. M, Praseeda. T.P. Phytosociological analysis of Mangroves at Kannur District, Kerala, Int. Journal of Env. Sci. 2011; 1(7): 671-677.

[10] MoFE. National wetland atlas: Kerala, Ministry of Forests and Environment, India 2010; 146p.

CHAPTER 22

Diversity of Stingless Bee Flora in Kerala

Devanesan S.*, Premila K.S. and Shailaja K.K.

AICRP on Honey bees and Pollinators, Vellayani Centre, Department of Agricultural Entomology, College of Agriculture, Kerala Agricultural University, Vellayani, Kerala, India

Abstract: The stingless bee plays a key role in pollination and yield enhancement of natural tropical plants and crops. The honey produced by stingless bee is unique and fetches high price because of its medicinal property. They collect nectar and pollen mainly from medicinal plants, low lying herbs and weeds having comparatively smaller flowers, which are not commonly visited by *Apis* spp. The rich biodiversity and abundant flora of the State is studied in relation to the foraging activity of stingless bees *Trigona iridipennis* Smith for its nectar and pollen. The study indicated that stingless bees visited 99 plants either for nectar or pollen or both. Out of these 99 plants recorded, 33 plants provided both nectar and pollen, 46 plants provided nectar only and 20 plants provided pollen alone.

Keywords: Flora, foraging plants, Kerala, meliponiculture, nectar, pollen, stingless bee, *Trigona iridipennis*.

1. INTRODUCTION

Kerala is having diversified flora which offers nectar and pollen to the stingless bee *Trigona iridipennis* Smith, which produce honey with great medicinal properties. They belong to the family Apidae and come under the sub-family Meliponinae. Stingless bees are so called because they have no venom apparatus (sting). They exhibit larger diversity than honey bees. The natural domicile is the crevices in walls, hollow blocks *etc*. Stingless bee honey is used worldwide as food and medicine. Despite the small amount of honey produced per hive, the honey is in great demand and fetches comparatively higher prices due to its medicinal value. Stingless bees are ecologically important because of their role in pollination of tropical plants and crops. Over 130 species of stingless bees' world over have been identified as potential pollinators of crops [1].

2. STINGLESS BEE FLORA IN KERALA

Preliminary study was conducted in random locations of each district of Kerala

***Corresponding author Devanesan S.:** Honey bees and Pollinators, Vellayani Centre, Department of Agricultural Entomology, College of Agriculture, Kerala Agricultural University, Vellayani, Kerala, India; Tel: 09847063300; Fax: 0471 2384422; E-mail: devanesanstephen@gmail.com

Laladhas K.P., Oommen V. Oommen and Sudhakaran P.R. (Eds.)

and the flowering plants were observed for stingless bee forage for a year. The survey was conducted through monthly visit to the selected sites. The flora visited by the stingless bees were observed from 6 am to 6 pm regularly in selected location of different districts during the blooming period and documented. Plants are visited by the bees either for nectar or pollen or both. Plants were scored as bee foraging species when at least three (3) stingless bees had visited and foraged on the flowers within the observation period of ten minutes. Whenever bees were found on the flowers of such plants, their foraging behaviour was observed and categorised as pollen gatherer based on the pollen load, on the pollen collecting legs and others as nectar gatherers. The floral and extra-floral nectaries of plants are visually observed and documented.

Ninety nine plants were recorded in which stingless bees visited for nectar or pollen or both. Of these 99 plants, 33 provided both nectar and pollen. They are categorized under medicinal plants (*Ixora coccinea*, *Ricinus communis*, *Azadirachta indica*, *Eucalyptus globulus)*, plantation crops (*Anacardium occidentale*, *Areca catechu, Camelia sinensis),* condiments and spices (*Tamarindus indica)*, vegetable crops (*Moringa oleifera*, *Raphanus sativus, Solanum melongena, Capsicum annum, Abelmoschus esculentus, Lycopersicon esculentum, Lagenaria vulgaris, Momordica charantia, Cucurbita moschata),* oil seeds (*Cocos nucifera, Helianthus annus),* fruit crops (*Psydium guajava*, *Litchi chinensis, Citrus* spp., *Artocarpus heterophyllus),* ornamental plants (*Rosa* spp., *Antigonon leptopus, Caesalpinia pulcherima, Samanea saman*, *Polianthus tuberosa)*, herbs, shrubs and bushes (*Manihot glaziovi)*, tuber crop (*Manihot esculenta),* forest plants (*Cauroupita guinensis*, *Muntingia calabura*, *Santalum album*).

Forty six plants provided nectar only. They are grouped under medicinal plants (*Indigofera tinctoria, Phyllanthus niruri, Aerva lanata, Baliospermum monatanum, Vinca rosea, Boerhaavia diffusa, Oxalis corniculata, Ocimum sanctum),* plantation crops (*Hevea brasiliensis*, *Coffea robusta, Coffea Arabica),* condiments and spices *(Elettaria cardamomi, Vanilla* sp., *Burmese coriander),* vegetable crops (*Brassica rapa pekinensis),* pulses (*Dolichos lablab, Vigna mungo, Cajanus cajan, Psophocarpus tetragonolobus),* oil seeds (*Sesamum indicum, Brassica juncea),* fruit crops *(Musa* spp., *Malphigia punicifolia)*, ornamental plants (*Impatiens balsaminae*, *Duranta goldiana, Hamelia patens, Tagetes erecta, Duranta plumieri, Canna indica, Euphorbia pulcherima, Heliconia rostrata, Gladiolus grandiflorus, Hibiscus* spp., *Datura* spp., *Dahlia spp., Etlinger elatior, Celosia argentea var. plumose),* herbs, shrubs and bushes *(Euphorbia hirta, Crotalaria verrucosa, Crotalaria mucronata, Sesbania rostrata,*

Jatropha, Cassia alata, Justicia simplex, Ziziphus nummularia), and green manure *(Gliricidia sepium).*

Pollen is provided by 20 plants and they are categorized under medicinal plants (*Mimosa pudica, Lowsonia alba),* condiments and spices *(Cinnamomum zeylanicum)*, vegetable crops (*Luffa cylindrica, Talinum triangulare, Averrhoa bilimbi, Allium cepa, Momordica cochinchinensis),* millets *(Pennisetum typhoides),* fibre crops (*Gossypium hirsutum, Corchorus olitorius, Bombax malabaricum)*, fruit crops (*Mangifera indica, Carica papaya),* and ornamental plants *(Anthurium andreanum, Solidago canadensis, Haemanthus cinnabarinus, Nymphea stellata, Salvia splendens, Calliandra surinamensis*).

Table 1: Stingless bee flora in Kerala

Common Name	Scientific Name	Family	Source
Medicinal Plants			
Indigo	*Indigofera tinctoria*	Papilionaceae	N
Touch-me-not	*Mimosa pudica*	Mimosaceae	P
Ixora	*Ixora coccinea*	Rubiaceae	N+P
Phyllanthus	*Phyllanthus niruri*	Euphorbiaceae	N
Castor	*Ricinus communis*	Euphorbiaceae	N+P
Javanese wool plant	*Aerva lanata*	Amaranthaceae	N
Nagadandhi	*Baliospermum monatanum*	Euphorbiaceae	N
Periwinkle	*Vinca rosea*	Apocynaceae	N
Boerhavia	*Boerhavia diffusa*	Nyctaginaceae	N
Creeping wood sorrel	*Oxalis corniculata*	Oxalidaceae	N
Henna	*Lowsonia alba*	Lythraceae	P
Holybasil (Thulsi)	*Ocimum sanctum*	Lamiaceae	N
Neem	*Azadirachta indica*	Meliaceae	N+P
Eucalyptus	*Eucalyptus globulus*	Myrtaceae	N+P
Plantation crops			
Rubber	*Hevea brasiliensis*	Euphorbiaceae	N
Cashew	*Anacardium occidentale*	Anacardiaceae	N+P
Coffee	*Coffea arabica*	Rubiaceae	N
Coffee	*Coffea robusta*	Rubiaceae	N
Areca nut	*Aereca catechu*	Arecaceae	N+P
Tea	*Camelia sinensis*	Theaceae	N+P
Condiments and spices			
Tamarind	*Tamarindus indica*	Fabaceae	N+P
Cinnamon	*Cinnamomum zeylanicum*	Lauraceae	P

Table 1: contd…

Cardamom	*Elettaria cardamomi*	Zingiberaceae	N
Vanilla	*Vanilla sp.*	Orchidaceae	N
Burmese	*Burmese coriander*	Umbelliferae	N
Vegetable crops			
Sponge gourd	*Luffa cylindrica*	Cucurbitaceae	P
Sweet gourd	*Momordica cochinchinensis*	Cucurbitaceae	P
Bottle gourd	*Lagenaria vulgaris*	Cucurbitaceae	N+P
Bitter gourd	*Momordica charantia*	Cucurbitaceae	N+P
Pumpkin	*Cucurbita moschata*	Cucurbitaceae	N+P
Drumstick	*Moringa oleifera*	Moringaceae	N+P
Chinese cabbage	*Brassica rapa pekinensis*	Umbelliferae	N
Waterleaf	*Talinum triangulare*	Portulacaceae	P
Radish	*Raphanus sativus*	Brassicaceae	N+P
Bilimbi	*Averrhoa bilimbi*	Oxalidaceae	P
Onion	*Allium cepa*	Amaryllidaceae	P
Brinjal	*Solanum melongena*	Solanaceae	N+P
Chilly	*Capsicum annum*	Solanaceae	N+P
Tomato	*Lycopersicon esculentum*	Solanaceae	N+P
Winged bean	*Psophocarpus tetragonolobus*	Fabaceae	N
Okra	*Abelmoschus esculentus*	Malvaceae	N+P
Pulses			
Dolichos	*Dolichos lablab*	Fabaceae	N
Redgram	*Cajanus cajan*	Fabaceae	N
Black gram	*Vigna mungo*	Fabaceae	N
Millets			
Bajra	*Pennisetum typhoides*	Poaceae	P
Oil seeds			
Coconut	*Cocos nucifera*	Palmaceae	N+P
Sunflower	*Helianthus annus*	Compositae	N+P
Gingely	*Sesamum indicum*	Pedaliaceae	N
Mustard	*Brassica juncea*	Umbelliferae	N
Fibre crops			
Cotton	*Gossypium hirsutum*	Malvaceae	P
Cotton tree	*Bombax malabaricum*	Malvaceae	P
Jute	*Corchorus olitorlus*	Tiliaceae	P
Fruit crops			P
Mango	*Mangifera indica*	Anacardiaceae	N
Banana	*Musa spp.*	Musaceae	N+P

Table 1: contd…

Guava	*Psydium guajava*	Myrtaceae	P
Papaya	*Carica papaya*	Caricaceae	N+P
Litchi	*Litchi chinensis*	Sapindaceae	N+P
Citrus	*Citrus spp.*	Rutaceae	N+P
Jack	*Artocarpus heterophyllus*	Moraceae	N
West Indian berry	*Malphigia punicifolia*	Malpighiaceae	
Ornamental plants			
Balsam	*Impatiens balsaminae*	Balsaminaceae	N
Rose	*Rosa* spp	Rosaceae	N+P
Coral creeper	*Antegonon leptopus*	Polygonaceae	N+P
Duranta	*Duranta goldiana*	Verbenaceae	N
Hamelia	*Hamelia patens*	Rubiaceae	N
Anthurium	*Anthurium andreanum*	Araceae	P
Marigold	*Tagetes erecta*	Compositae	N
Golden dew drop	*Duranta plumieri*	Verbenaceae	N
Canna	*Canna indica*	Cannaceae	N
Poinsettia	*Euphorbia pulcherima*	Euphorbiaceae	N
Golden rod	*Solidago canadensis*	Compositae	P
Ball lilly	*Haemanthus cinnabarinus*	Amaryllidaceae	P
Water lily	*Nymphea stellata*	Nymphaeaceae	P
Sage	*Salvia splendens*	Lamiaceae	P
Gladiolus	*Gladiolus grandiflorus*	Iridaceae	N
Shoe flower	*Hibiscus* spp	Malvaceae	N
Angel's trumpet	*Datura* spp	Solanaceae	N
Dahlia	*Dahlia* spp	Asteraceae	N
Ginger lily	*Etlinger elatior*	Zingiberaceae	N
Bird of paradise	*Heliconia rostrata*	Zingiberaceae	N
Cocks comb	*Celosia argentea*	Amaranthaceae	N
Powder puff	*Calliandra surinamensis*	Fabaceae	P
Peacock plant	*Caesalpinia pulcherima*	Fabaceae	N+P
Rain tree	*Samanea saman*	Fabaceae	N+P
Tube rose	*Polianthes tuberosa*	Asparagaceae	N+P
Herbs, shrubs and bushes			
Asthma weed	*Euphorbia hirta*	Euphorbiaceae	N
Wild tapioca (Serra rubber)	*Manihot glaziovi*	Euphorbiaceae	N+P
Jatropha	*Jatropha sp.*	Euphorbiaceae	N
Crotolaria	*Crotalaria verrucosa*	Fabaceae	N
Crotalaria	*Crotalaria mucronata*	Fabaceae	N

Table 1: contd...

Sesbania	*Sesbania rostrata*	Fabaceae	N
Cassia	*Cassia alata*	Fabaceae	N
Justicia	*Justicia simplex*	Acanthaceae	N
Ziziphus	*Ziziphus nummularia*	Rhamnaceae	N
Tuber crops			
Tapioca	*Manihot esculenta*	Euphorbiaceae	N+P
Green manure crop			
Gliricidia	*Gliricidia sepium*	Fabaceae	N
Forest Trees			
Cannon ball tree	*Couroupita guianensis*	Lecythidaceae	N+P
Birds cherry	*Muntingia calabura*	Muntingiaceae	N+P
Sandal	*Santalum album*	Santalaceae	N+P

N - Nectar P - Pollen

Ninety nine plants were recorded as stingless bee flora in the present study. Of this cardamom [2], sunflower [3], sesamum, onion [4], mango [5], delonix, neem, ageratum, bamboo, antigonon, eucalyptus [6] cotton [7], salvia and lavender [8], orchid [9], coconut [10, 11] and caesalpinia, euphorbia [12] were reported earlier. In addition rubber, cashew, chilly, cannon ball tree, drumstick, guava, bird's cherry, tamarind, anthurium, bajra, sponge gourd, cinnamon, ixora, touch-me-not, balsam, banana, duranta, hamelia, marigold and phyllanthus [13] was also reported as stingless bee foraging plants.

The presence of this much of foraging plants confirms the scope for commercial meliponiculture in Kerala. Conservation of this flora is important for sustainable Meliponiculture resulting in enhanced stingless bee honey production and in turn pollination and yield enhancement for food security.

ACKNOWLEDGEMENTS

Declared none.

CONFLICT OF INTEREST

The authors confirm that this chapter contents have no conflict of interest.

REFERENCES

[1] Roubik DW, Eds. Stingless bee colonies for pollination. FAO (UN), Rome 1995; pp. 324.

[2] Puttanshetty, HV, Prasad ABN. Foraging behaviour of honeybees on cardamom (*Elettaria cardomomum* Malton). Curr. Res. 1973; 2: 56.

[3] Goel SC, Kumar A. Population aggregate of *Melipona iridipennis* Smith. (Apidae) on sunflower. Indian J. Hort. 1981; 38: 125-128.

[4] Mohan RG, Lazar M, Suryanarayana MC. Foraging behaviour of honey bees in sesame (*Sesamum indicum* L.). Indian Bee J. 1981; 43: 97-100.

[5] Singh G. Insect pollinators of mango and their role in fruit setting. Acta Hort. 1989; 231: 629-632.

[6] Ramanujam CGK, Fatima K, Kalpana TP. Nectar and pollen sources for Dammer Bee (*Trigona iridipennis* Smith) in Hyderabad, India. Indian Bee J. 1993; 55: 25-28.

[7] Mohan RG. Studies on floral biology and foraging behaviour of honey bees on coconut palm, *Cocos nucifera* L. Indian Bee J. 1996; 59: 238-239.

[8] Mohan RG, Singh TSMS. Foraging behaviour of honey bees on salvia. Indian Bee J. 1997; 59: 2.

[9] Roubik DW. Potential pollination out crossing distances flown by honey bees in a Congo forest. J. Kans. Ent. Soc. 2000; 73: 137-150.

[10] Devanesan S, Nisha MM, Bennet R, Shailaja KK. Foraging behaviour of stingless bee *Trigona iridipennis* Smith. Insect Environment 2002; 8(3): 131.

[11] Muthuraman M, Suresh Desai D, Thangavelu GK, Srinivasan G. Coconut inflorescences Sap: A food source for honeybees. Insect Environ. 2002; 7: 4-5.

[12] Premila KS, Bennet R, Shailaja KK, Devanesan S. Flora for stingless bee, *Trigona iridipennis* in Kerala. International Workshop on Conservation and Management of bees for sustainable development in honey festival (APIEXPO –2003) 13-18 Oct. 2003 Bangalore 2003; pp. 22.

[13] Devanesan S, Shailaja KK, Premila KS. Status paper on Stingless bees. AICRP on Honey bees and Pollinators, ICAR, New Delhi. 2009; pp. 80.

Chapter 23: Seasonal Variations Influencing the Bryophyte Diversity of Monoculture Plantations in the Southern Western Ghats

CHAPTER 23

Seasonal Variations Influencing the Bryophyte Diversity of Monoculture Plantations in the Southern Western Ghats

Kariyappa K.C. and Daniels A.E.D.*

Bryology Laboratory, Department of Botany & Research Centre, Scott Christian College, Nagercoil, Tamil Nadu, India

Abstract: Two monoculture plantations *viz.*, Rubber and Teak were sampled over a period of 10 months from August 2009 to June 2010 for Bryophytes. A total of 1315 collections were made from 30 permanent quadrates laid on 6 transects. Rubber plantation was found to harbour 14 species and Teak 22. The species diversity increased during monsoon season and declined in drier months. This pattern was more pronounced in Teak plantation. On the other hand species found during the drier months are the more generalized and light tolerant mosses such as *Fissidens ceylonensis, F. kalimpongensis, Hyophila comosa, H. involuta* and liverworts such as *Frullania acutiloba* and *F. campanulata*. However, a significant variation in the number and type of species was not seen in both the plantations. This indicates that the monoculture plantations remain stable and promote the growth of more generalised and light tolerant species but lead to local extinction of species with restricted distribution.

Keywords: Bryophytes, corticolous, diversity, evergreen forests, humidity, light tolerant, monoculture plantations, rubber, rupicolous, seasonal changes, shade loving, teak, terricolous.

1. INTRODUCTION

Bryophytes are the simplest and the most primitive of land plants lacking a well developed vascular tissue system. Their adaptation to a terrestrial mode of life is partial as water is indispensable at one stage or other in their life cycle. Hence, they are more common in humid areas and during rainy seasons. Bryophytes prefer microclimatic niches such as crevices of rocks and trees, and vicinity of shady springs and streams. Many are epiphytes growing on bark of trees (corticolous), leaves (epiphyllous or folicolous), rocks (rupicolous), pebbles and stones (saxicolous), logs (lignicolous), river banks and roadside cuts (terricolous).

***Corresponding author Daniels A.E.D.:** Bryology Laboratory, Department of Botany & Research Centre, Scott Christian College, Nagercoil, Tamil Nadu, India; Tel: 04652-279031; Fax: 04652-229800; E-mail: dulipdaniels@yahoo.co.uk

Laladhas K.P., Oommen V. Oommen and Sudhakaran P.R. (Eds.)

A study of the diversity of bryophytes along a gradient of human disturbance in the Southern Western Ghats showed that evergreen forests contain the highest bryophyte diversity followed by degraded evergreen forests and monoculture plantations [1]. Moreover, evergreen forests harbour the highest number of exclusive species whereas plantations promote the growth of more generalised and light tolerant ones. Presently, the influence of seasonal variations on the bryophyte diversity in two monoculture plantations in the Southern Western Ghats has been studied.

2. METHODS

Two monoculture plantations *viz*., rubber and teak at 200 m altitude were selected on the hills of Lower Kodaiyar in the Southern Western Ghats of Kanyakumari District. Sampling was carried out over a period of 11 months from August 2009 to June 2010. Three fixed distance transects of 50 m length with five 10 x 10 m quadrates alternating with each other were laid in each plantation. Altitude, light intensity, relative humidity, temperature and rainfall were recorded using standard branded instruments. Transects were laid permanently in such a way that each quadrate on the transect was visited and sampled once every month for 10 months. Collections and identifications [2] from all habitats found in the plantations such as bark of trees, rocks, stones and soil were made. The diversity of the plantations was assessed using Shannon-Weiner Index [3, 4]. Correlation coefficient (MS office software package 2003) was used to correlate environmental parameters with the bryophyte diversity.

3. RESULTS AND DISCUSSION

On the whole 1315 collections were made from 30 permanent quadrates laid on 6 transects. A total of 668 and 647 colonies were collected from rubber and teak plantations respectively. Rubber plantation hosts 6 genera and 14 species (Table **3**) whereas teak plantation hosts 12 genera and 22 species (Table **4**). Rubber showed the highest diversity (H′ = 1.0015) in the month of August and teak (H′ = 1.244) in the month of October (Table **1**). The comparatively lower diversity in rubber plantation may be due to the regular intercultural practices like weeding, manuring and use of pesticides and fungicides. However, from Table **1**, it is evident that in both monoculture plantations the species diversity remains more or less stable. From January a drop in the diversity can be seen in both and more so in teak. Rubber being an evergreen species basically remains green all the year unless monsoon fails. But teak being a deciduous species leaves its canopy open due to leaf fall thereby permitting sunlight to reach the floor of the plantation which eliminates most of the species since bryophytes are moisture and shade loving plants. This is evident from Tables **1** and **2** which show higher diversity during monsoon season and the average of the parameters correlated to the bryophyte diversity in each plantation respectively. Light and temperature show a negative correlation which is true in the case of bryophytes since higher light intensity raises the temperature which ultimately lowers the bryophyte diversity. This speaks beyond doubt that bryophytes are shade loving plants [5]. Though humidity appears to be insignificant in influencing the bryophyte diversity in this study, higher the humidity greater is the diversity [5]. The reason for such a low humidity may be due to the very open conditions prevailing in the plantations other than the open canopy during non-monsoon seasons.

Table 1: Higher diversity during monsoon season

Month	H′ Rubber	H′ Teak
August 2009	1.0015	0.71
September 2009	0.935	1.15
October 2009	0.917	1.244
November 2009	0.935	1.18
December 2009	0.902	1.06
January 2010	0.695	0.644
February 2010	0.782	0.721
March 2010	0.658	0.484
April 2010	0.635	0.391
May 2010	0.717	0.398
June 2010	0.812	0.492

Table 2: Correlation between bryophyte diversity and environmental parameters

Parameters	Rubber	Teak
Humidity	0.4741	0.3261
Light (Lux)	-0.70805*	-0.7627*
Rainfall	0.244617	0.728279*
Temperature	-0.7693*	-0.7603*

*Significant at 5% level.

Table 3: Habitat diversity in the Rubber plantation

Sl. No.	Species	Corticolous	Rupicolous	Terricolous
1	*Archilejeunea apiculifolia*	+	-	-
2	*Archilejeunea minutilobula*	+	-	-
3	*Fissidens auriculatus*	-	-	+
4	*F. ceylonensis*	-	+	+
5	*F. kalimpongensis*	-	-	+
6	*F. zollingeri*	-	-	+
7	*F. zippellianus*	-	-	+
8	*Frullania acutiloba*	+	-	-
9	*F. campanulata*	+	-	-
10	*F. muscicola*	+	-	-
11	*Hyophila comosa*	-	+	+
12	*H. involuta*	-	+	+
13	*Isopterygium lignicola*	-	-	+
14	*Mastigolejeunea auriculata*	+	-	-

From the two plantations only 7 species of liverworts have been recorded *in situ.*, *Archilejeunea apiculifolia, A. minutilobula, Fossombronia himalayensis, Frullania acutiloba, F. campanulata, F. muscicola* and *Mastigolejeunea auriculata*. Of these, only *Fossombronia himalayensis* is a terricolous form. The rest are either corticolous or rupicolous forms. In the plantations studied they occur only as corticolous forms. This is a clear indication that plantations possess low humidity because leafy liverworts generally prefer to grow as corticolous forms in low humidity areas. Moreover, these corticolous liverworts are the most common ones occurring in degraded forests and clove plantations [1].

Of the 19 mosses recorded in the plantations *Fissidens ceylonensis, F. zollingeri, Hyophila comosa, H. involuta, Philonotis fontana* and *P. thwaitesii* are the most common light tolerant species. Moreover, all the moss species recorded occur either as rupicolous or terricolous forms except *Thuidium cymbifolium* (Table **4**). This is a clear indication that all the moss species except *Thuidium cymbifolium* are light tolerant ones.

Table 4: Habitat diversity in the Teak plantation

Sl. No.	Species	Corticolous	Rupicolous	Terricolous
1	*Bryum porphyroneuron*	-	+	+
2	*Barbula vinealis*	-	-	+
3	*Frullania acutiloba*	+	-	-
4	*F. campanulata*	+	-	-
5	*Fissidens ceylonensis*	-	+	+
6	*F. involutus*	-	-	+
7	*F. kalimpongensis*	-	-	+
8	*F. pokhrensis*	-	-	+
9	*F. splachnobryoides*	-	-	+
10	*F. subpalmatus*	-	-	+
11	*F. zollingeri*	-	-	+
12	*F. zippellianus*	+	-	-
13	*Fossombronia himalayensis*	-	-	+
14	*Hyophila comosa*	-	+	+
15	*H. involuta*	-	+	+
16	*Isopterygium lignicola*	-	-	+
17	*Pohlia longicolla*	-	-	+
18	*Philonotis fontana*	-	-	+
19	*P. thwaitesii*	-	-	+
20	*Stereophyllum confusum*	-	+	-
21	*Taxithelium nepalense*	-	-	+
22	*Thuidium cymbifolium*	+	-	-

CONCLUSION

From the present study it is evident that monoculture plantations do not harbour rare and delicate bryophyte species but rather more generalised and light tolerant ones. Moreover, these plantations appear to remain more stable and resist seasonal changes that may ultimately lead to the gradual elimination of delicate shade loving species.

ACKNOWLEDGEMENTS

Declared none.

CONFLICT OF INTEREST

The authors confirm that this chapter contents have no conflict of interest.

REFERENCES

[1] Daniels AED, Kariyappa KC. Bryophyte diversity along a gradient of human disturbance in the southern Western Ghats. Curr. Sci. 2007; 93: 976-982.

[2] Gangulee HC. Mosses of Eastern India and adjacent regions. 1 – 3 (Fasc. 1-8). Calcutta: Author. 1969-1980.

[3] Daniels RJR, Hegde M., Gadgil M. Birds of the man-made ecosystem: The plantations. Proc. Indian Acad. Sci. (Anim. Sci.) 1990; 99: 79-89.

[4] Pielou. EC. Ecological diversity, John Wiley, New York 1975.

[5] Marshall M, Proctor MCF. Are Bryophytes shade plants? Photosynthetic light responses and proportions of Chlorophyll a, Chlorophyll b and total carotenoids. Ann. Bot. 2004; 94: 593-603.

CHAPTER 24

Macroinvertebrates Associated with Aquatic Macrophytes *Monochoria vaginalis* and *Eichhornia crassipes* in Ponnani Kole

Sanila V.K. and Razia Beevi M.*

Department of Aquaculture and Fishery Microbiology, M.E.S. Ponnani College, Ponnani Malappuram Dt., Kerala, India

Abstract: The Kole land supports a variety of macrophytes, and their associated fauna. The present study was conducted to assess the macroinvertebrate communities associated with the microhabitats of macrophytes, *Monochoria vaginalis* and *Eichhornia crassipes* from the Ponnani kole. The samples were collected during pre-monsoon, monsoon and post monsoon seasons from 1 m^2 area of the vegetation. The area was demarked into quadrates. The samples were then taken using a scoop net and immediately transferred in to a polythene bag along with water. The vegetation in the polythene bag which served as the sampling unit was taken to the laboratory to observe the abundance and diversity of macroinvertebrates associated with the macrophytes. Physico-chemical parameters such as pH, temperature, salinity and dissolved oxygen were also analyzed. The number of organisms in each group associated with the macrophytes were identified. The identified taxa were Annelida, Crustacea, Insecta, Arachnida, Gastropoda and Bivalvia. The taxa Insecta & Crustacea were found dominant in both *Monochoria vaginalis* and *Eichhornia crassipes* during pre and post monsoon. While during monsoon Insecta alone were found dominating in both the macrophytes. Twenty seven species of macroinvertebrates were found associated with the two macrophytes. On *Monochoria vaginalis* a total of nineteen species and on *Eichhornia crassipes* twenty two species of macroinvertebrates were observed. The results of the present study reveal that the two macrophytes provide habitat for diverse invertebrates, the abundance of which denotes the rich productivity of the area.

Keywords: *Eichhornia crassipes,* fauna, kole land, macroinvertebrates, macrophytes, *Monochoria vaginalis*, wetland ecosystem.

1. INTRODUCTION

The kole land, one of the rice granaries of Kerala is a unique wetland ecosystem down south of India. The kole land runs parallel to the sea and are low lying tracts

***Corresponding author Razia Beevi M.:** Department of Aquaculture and Fishery Microbiology, M.E.S. Ponnani College, Ponnani Malappuram Dt., Kerala, India; Tel: 9446558046; E-mail: razia.mes@gmail.com

Laladhas K.P., Oommen V. Oommen and Sudhakaran P.R. (Eds.)

located 0.5 to 1 meter below the sea level, extending to an area of 13632 ha, spread over Thrissur and Malalappuram districts of Kerala. The kole land in Thrissur is located in Mukundhapuram, Thrissur and Chavakkad taluks, where as Ponnani kole is located in Chavakkad and Thalapilly taluks of Thrissur district and ponnani taluk of Malapuram district. Macrophytes are important primary producers along the coast worldwide, serving as habitat or functioning as ecological engineering species. Diverse invertebrate communities exist among aquatic vegetation of kole lands. Invertebrates utilize macrophytes as a direct food source [1], shelter from predators, spawning and attachment sites as well as feeding on the periphyton growing on their surfaces. In this study the associated invertebrates of the macrophytes *'Eichhornia crassipes'* and *'Monochoria vaginalis'* under the same family Pontederaceae were examined.

2. MACRO INVERTEBRATES

Macrophytes were collected from 1 m^2 area of the vegetation from the lower stretches of Ponnani kole during pre monsoon, monsoon and post monsoon. Identification of specimens was carried out down to the lowest possible level, except the aquatic worms which were identified as Oligochaeta. The specimens were identified with the help of identification key [2]. Macro invertebrates recorded during present study were shown in Table **2**. The identified taxa were Annelida, Crustacea, Arachnida, Insecta Gastropoda and Bivalvia. A total of twenty seven species of organisms were collected.

During pre monsoon, on *Monochoria vaginalis* arachnida was more in number, while that on *Eichhornia crassipes* was crustacea. The class insecta was the most important in terms of species abundance on both *Monochoria vaginalis* and *Eichhornia crassipes*. The other numerically important classes among the fauna on *Monochoria* were gastropoda represented by *Planobis corneous.* On *Eichhornia* it was annelida, gastropoda, crustacea, and oligochaeta represented respectively by *Haemadipsa cochiniana, Planorbis corneus, Fenneropenaeus indicus,* and aquatic worms.

During monsoon, the numerically most important class was insecta on both the plants. The same class insecta was the most important in terms of species abundance on *Monochoria* represented by two species. On *Eichhornia* it was insecta and annelida equally contributing three species. The other classes in the fauna were oligochaeta, crustacea and arachnida represented by aquatic worms, *Macrobrachium* sp. and *Araena* sp. respectively on *Monochoria.* On *Eichhornia* there were only class crustacea represented by *Caridina typus* and *Asellus* sp.

During post monsoon, on *Monochoria* insecta was the numerically most important faunal community, represented equally by *Amphizoa* sp. and Predacious diving beetle while that of *Eichhornia* was Crustacea represented by *Macrobrachium* sp. The class Insecta was the most important in terms of species abundance on both the plants represented by nine and seven species respectively. The other numerically important classes in the fauna on *Monochoria* were Arachnida, Oligochaeta, Gastropoda, and Bivalvia represented respectively by *Araena* sp, Aquatic worms, *Planorbis corneus, Lamellidens marginalis*. On *Eichhornia* it was Annelida, Insecta, Gastropoda, and Arachnida, represented by *Hirudinaria granulosa, Pentaneura monilis, Planorbis corneus,* and *Araena* sp.

On *Monochoria vaginalis* a total of nineteen species and on *Eichhornia crassipes* twenty two species of macro invertebrates were found (Table **2**). Seasonal analysis of macro invertebrates showed high species diversity during post monsoon followed by pre monsoon and the monsoon (Table **1**). On *Eichhornia crassipes* the species diversity is high during post monsoon and declined during pre monsoon and monsoon (Fig. **2**). While in *Monochoria vaginalis* the number of species found were almost same during post monsoon and pre monsoon and showed a decline during the monsoon (Fig. **1**)

Table 1: Abundance of Macro Invertebrates on *Monochoria vaginalis* (Plant- I) and *Eichhornia crassipes* (Plant- II)

	Invertebrate Species	**Pre Monsoon**		**Monsoon**		**Post Monsoon**	
		Plant-I	**Plant-II**	**Plant-I**	**Plant-II**	**Plant-I**	**Plant-II**
1	*Haemadipsa cochiniana*		7		1		2
2	*Hirudinaria granulosa*		2		2		7
3	*Glossiphonia complanata*		1		1		
4	*Lamellidens marginalis*	1				2	
5	*Planorbis corneus*	4	6			5	4
6	*Melanoides tuberculata*	1				2	
7	*Araena* sp.(spider)	5	3	1		7	3
8	*Asellus* sp.(Aquatic sow bug)				1		
9	*Fenneropenaeus indicus*					2	6
10	*Caridina typus*				3		3
11	*Macrobrachium* sp.		8				9
12	Oligochaets	1	2	2		5	
13	*Belostoma indicum*			2		1	
14	*Dragon fly*- nymph		1		2	2	1

Table 1: contd…

15	*Damsel fly*- nymph					2	1
16	*Caddis fly*- larvae	1		1	1	2	1
17	*Pentaneura monilis- larva*						6
18	*Chironomous tentans*-larva		1				
19	*Cybister* -larva					1	
20	*Leptocorisa acuta*	1					
21	Snout beetle	1					2
22	Predacious diving beetle	3		7	5	8	3
23	*Amphizoa* sp.		1			8	
24	Whirling beetle		1				
	3 unidentified insects						
25	a		1			3	1
26	b	2					
27	c		6			1	
Total number of species (species diversity)		**10**	**13**	**5**	**8**	**15**	**14**

Table 2: Macroinvertebrates associated with macrophytes *Monochoria* vaginalis and *Eichhornia crassipes*

Sl. No.	Class	*Monochoria vaginalis*	*Eichhornia crassipes*
1	Annelida	Aquatic worms	*Haemadipsa cochiniana* *Hirudinaria granulosa Glossiphonia complanata* Aquatic worms
2	Gastropoda	*Planorbis corneus* *Melanoides tuberculata*	*Planorbis corneus*
3	Bivalvia	*Lamellidens marginalis*	
4	Arachnida	*Araena sp.*	*Araena sp.*
5	Crustacea	*Fenneropenaeus indicus* *Macrobrachium sp.*	*Asellus sp. (Aquatic sow bug)* *Caridina typus* *Fenneropenaeus indicus Macrobrachium sp.*
6	Insecta	*Belostoma indicum* Dragon fly- nymph Damsel fly- nymph Caddis fly- larvae *Cybister* –larva *Leptocorisa acuta*	Dragon fly- nymph Damsel fly- nymph Caddis fly- larvae *Pentaneura monilis* - larvae of Tanypodinae *Chironomous tentans*- larve of

Table 2: contd...

		Snout beetle *Amphizoa* sp.- crawling beetle Predacious diving beetle	Chironomid Predacious diving beetle Snout beetle Whirling beetle
		Unidentified- 3	Unidentified-2

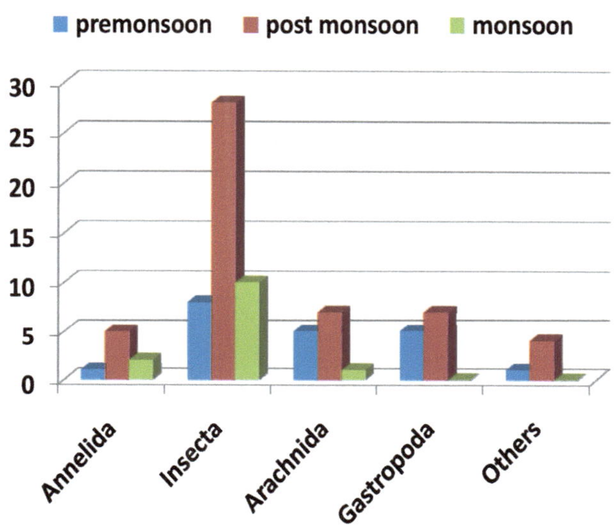

Figure 1: Abundance of macroinvertebrates in *Monochoria vaginalis.*

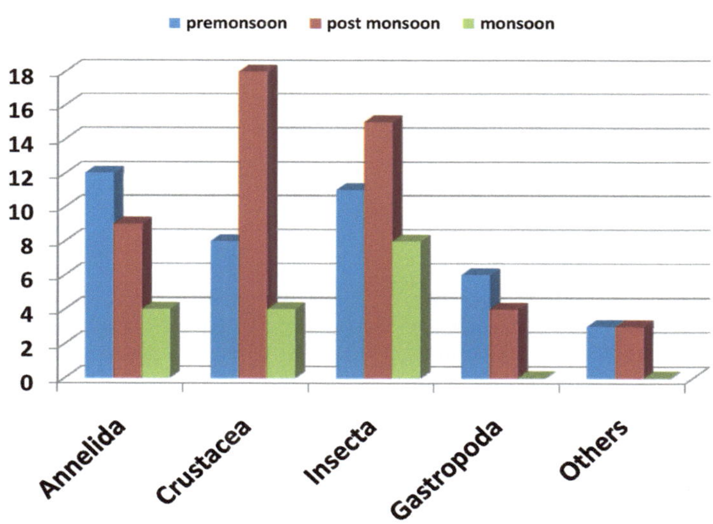

Figure 2: Abundance of macroinvertebrates in *Eichhornia crassipes.*

The class Annelida represents four species on *Eichhornia crassipes* and only one species on *Monochoria vaginalis*. The class Gastropoda represents two species on *Monochoria* and only one on *Eichhornia*. Bivalvia represents one species which is found only on *Monochoria vaginalis*. Arachnida seemed to favour both the plants. Crustacea includes four species, of which two were found on both the plants. Class insecta was the most abundant and diverse group which was represented by twelve species on *Monochoria vaginalis* and ten species on *Eichhornia crassipes*. Among Insects *Amphizoa* sp. Cybister larvae, *Leptocoriza acuta* clearly had a preference for *Monochoria vaginalis* and larvae of *Pentaneura monilis, Chironomous tentans,* and Whirling beetle on *Eichhornia crassipes*. Nymphs of Dragon fly, Damsel fly, Larvae of Caddis fly, Snout beetle, predacious diving beetle were seen in both the plants. In the case of Crustacea, *Asellus* sp. and juveniles of *Caridina typus* were found only on *Eichhornia crassipes* and juveniles of *Fenneropenaeus indicus* and *Macrobrachium* sp. were seemed to favour both *Monochoria vaginalis* and *Eichhornia crassipes*.

The rich faunal assemblage in both *Eichhornia crassipes* and *Monochoria vaginalis* showed the greater productivity of the kole land. The kole lands are being disturbed by land reclamation and habitat destruction due to various anthropogenic interventions, this area is to be protected to enhance the fish production and also for conserving the biodiversity of the unique ecosystem.

ACKNOWLEDGEMENTS

Declared none.

CONFLICT OF INTEREST

The author(s) confirm that this chapter contents have no conflict of interest.

REFERENCES

[1] Gregg WW, Rose FL. Influences of aquatic macrophytes on invertebrate community structure, guild structure, and microdistribution in streams. Hydrobiologia 1985; 128: 45-56.

[2] Jayasree DM, Roy SP, Datta Munshi JS. Manual of fresh water biota. Narendra publishing house, Delhi 2010.

CHAPTER 25

Micropropagation of an Exquisite Orchid, *Acampe rigida* (Buch. – Ham. Ex J. E. Sm.) P.F. Hunt., Through Embryo Culture

Gangaprasad A., Anjusha S. and Sibin N.P.

Plant Tissue Culture and Molecular Biology Lab, Department of Botany, University of Kerala, Kariyavattom, Thiruvananthapuram, Kerala, India

Abstract: Orchids consists of biologically interesting and commercially significant groups of plants representing the largest flowering plant family with nearly 30,000 species in 800 genera; about 1300 species occur in India. Green pod culture of *Acampe rigida*, an exquisite orchid of Western Ghats was tested in four different culture media *in situ*. Mitra *et al.*, Knudson C, Vacin and Went and Wimber media. Mitra *et al.*, medium proved to be the best for obtaining high percentage seed germination. Green pod culture of this species was further raised with ease in Mitra *et al.*, liquid medium fortified with different organic additives like casein acid hydrolysate (CH), peptone (P), yeast extract (YE), and coconut water (CW). Mitra *et al.*, nutrient medium supplemented with 0.05% CH (w/v) to achieve maximum seed germination and protocorm development. CW was found be least effective in seed germination and subsequent protocorm development. After 60 days of culture in liquid medium, the developing protocorms were transferred to solid Mitra medium containing same amount of additives. Maximum proliferation of protocorm was achieved in CH containing medium. After second subculture in solid medium, the healthy rooted seedlings obtained were transferred to community pots containing charcoal and tile pieces and 83% survival rate was obtained.

Keywords: *Acampe rigida*, embryo culture, green pod culture, liquid medium, micropropagation, orchid, protocorm, seedlings, seed germination, survival, tissue culture.

INTRODUCTION

The Sahyadri Mountain ranges of the Western Ghats region is an area of high plant genetic diversity and endemism [1]. The forests of this region are a rich repository of tropical flora and an abode of orchids. Western Ghats one of the

***Corresponding author Gangaprasad A.:** Plant Tissue Culture and Molecular Biology Lab, Department of Botany, University of Kerala, Kariyavattom, Thiruvananthapuram, Kerala, India; Tel: 09447552783; Fax: 04712308301; E-mail: agangaprasad@yahoo.com

Laladhas K.P., Oommen V. Oommen and Sudhakaran P.R. (Eds.)

richest floristic regions in the country is having about 4000 flowering plant species, about 30% of the flowering plants in the country [2].

Orchids are among the world's most diverse and widely distributed plants. They have complex life cycle, mycorrhizal association and specific pollination syndrome. It is a family of considerable economic importance particularly in horticulture and floristry. Many orchid species are threatened globally by over-collection from natural habitat for horticultural purposes [3]. The major threat to orchid populations is human activities such as land clearing for agriculture purpose, and urban development and collection of plant for medicinal and other purposes. Many orchids are rich in alkaloids and other phytochemical contents and many of them are effectively used to cure a wide range of ailments.

Plant tissue culture and micropropagation technique play an important role in conservation programme and management of botanical collection. Plant tissue culture is an effective tool to conserve plant genes and guarantee the survival of endangered and overexploited genotypes. For orchids, seed (embryo) cultures offer opportunities of easy and rapid multiplication, maintenance of diversity within species and facilitate reintroduction and restoration of taxa into natural habitat. The Ministry of Environment and Forest, Government of India and *International Union for Conservation of Nature* (IUCN) have recommended application of seed and tissue culture to save and multiply endangered taxa and their conservation through reintroduction. A cursory of the available literature reveals that conservation of rare, endemic and exquisite taxa has seldom been the subject of earlier investigations.

Acampe, a genus of about 15 species belongs to the family Orchidaceae is mostly distributed over Asiatic and African tropics. *A. rigida* is an exquisite epiphytic herb distributed in Pathanamthitta and Thiruvananthapuram districts of Kerala. Geographically it is distributed from Africa to Indo –Malaysia. *A. rigida* is a robust plant, stem woody and 30-40 cm long, leaves fleshy, coriaceous and 10-15cm long (Fig. **1**). Sepals and petals are yellowish barred transversely with red; lip white with a greenish yellow spur.

TISSUE CULTURE

Unripened fruits (green capsules) of *A. rigida* collected approximately 6 months after pollination was used for embryo culture experiments. The green capsules at the time of harvest were 5.9 cm in length, 3.4 cm width (Fig. **2**). Microscopic assessment revealed 78% viable seeds (Fig. **3**). After surface sterilization uniform seed suspensions were inoculated into presterilised Knudson, [4, 5], basal liquid nutrient medium supplemented with 2% sucrose. The inoculations were done under sterile conditions. All the cultures were incubated at 25±2˚C under 12hr photoperiod using cool – white fluorescent tubes and with constant agitation at 80rpm (Orbitek gyratory shaker). Observations of seed germination were made at regular intervals. After 60 days of culture, samples of developing protocorm were collected for determination of size and fresh weight. After 60 days, the developing protocorm were transferred to Mitra *et al.*, agar (0.8%) medium containing basal and same amount of different additives *viz.* CH, P, YE and CW. The proliferated protocorms with shoot initials were again transferred to fresh medium. After 60 days in the second subculture, healthy seedlings were deflasked, washed in tap

Figure 1: *Acampe rigida* habit – flower close up.

Figure 2: Green capsule of *Acampe rigida*

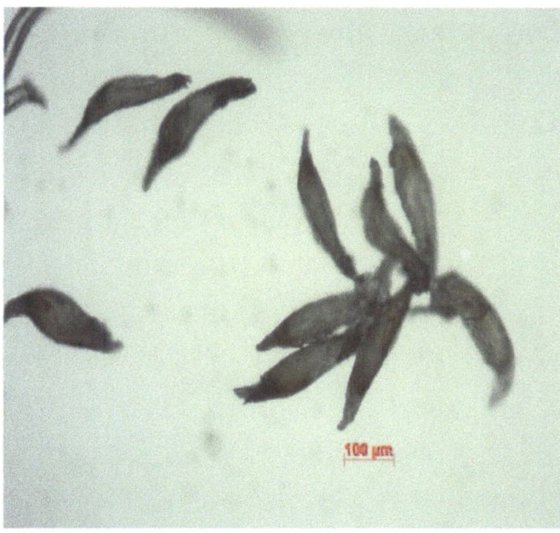

Figure 3: Microscopic view of seeds of *A. rigida.*

Figure 4: Protocorm development in Mitra *et al.*, liquid medium containing 0.05% CH.

Figure 5: Protocorm proliferation in Mitra *et al*., solid medium containing 0.05% CH.

Figure 6: Rooted seedlings of *A. rigida*.

water to remove traces of agar particle and dipped in 1% Indofil M 45 (fungicide) solution. The seedlings with 2-4 leaves and 3-5 roots were planted in clay pots containing charcoal and tile pieces (1:1). The potted plants were maintained in a nursery under diffuse sunlight and watered once a day.

The immature embryos of *A. rigida* transferred to four different basal nutrient media *viz*. Mitra *et al*., Vacin and Went, Wimber and Knudson C media responded positively. But maximum percentage of seed germination was obtained in Mitra *et al*., media (Table **1**). Therefore further experiments were conducted in Mitra *et al*., medium fortified with various organic additives. In all the media tried, the swelling and glistering of embryo were noticed after two weeks of inoculation and in most cases the embryo turned green yellow in colour. Of the different organic additives tested maximum percentage of seed germination was noticed in 0.05% CH (94%) followed by yeast extract and peptone (Table **2**, Fig. **4**). CW was found be least effective with (68%) of seed germination. CH containing medium promoted maximum protocorm growth as estimated from the diameter and fresh weight of the protocorm. An average of 2.08 mm diameter protocorm with 8.75 mg/ protocorm fresh weight was observed (Table **2**). Both peptone and YE had only marginal effect. Poor seed germination (68%) and protocorm growth (1.16mm diameter and 3.18mg/protocorm) was observed in 20% CW but it was superior to Mitra *et al*., basal medium (Table **2**). Maximum protocorm proliferation (daughter protocorm) was observed in CH containing medium. An average of 5.67 daughter protocorm was obtained from a single protocorm after the second subculture (Fig. **5**). Protocorm proliferation was also noticed in YE and P containing medium but number was less than that of CH (Table **3**). There was no protocorm proliferation in basal and CW containing medium. First leaf development was observed after 45 days of first subculture in solid medium. Seedlings with well developed roots were formed after the second subculture (Fig. **6**). Under nursery condition about 83 % seedling survived upon transfer to community pots.

Table 1: Green pod culture of *Acampe rigida* in different liquid basal nutrient medium

Media	Germination (%)	Pigmentation
Mitra	84	Yellowish green
Knudson C	68	Yellowish green
Vacin & Went	54	Light yellow
Wimber	57	Yellow

Observation were made after 60 days of inoculation.

Table 2: Effect of different additives on seed germination in *Acampe rigida* on Mitra *et al.*, medium

Additives	Germination (%)	Colour of the protocorm	Response after 60 days	
			Fresh wt (mg) of protocorm	Diameter of protocorm
Basal	84	Yellowish green	2.69±0.22	1.42±0.21
0.05% CH	94	Greenish yellow	8.75±2.08	2.08±0.21
0.05% P	87	Yellowish green	7.00±1.21	1.65±0.23
0.05% YE	88	Yellowish green	6.31±1.14	1.74±0.28
20% CW	68	Yellow	3.18±0.27	1.16±0.17

Table 3: Effect of different organic additives on protocorm proliferation in *Acampe rigida*

Medium	Additives	First subculture	Second subculture
Mitra *et al.*,	NIL	1.27±0.59	1.33±0.82
	0.05% CH	4.80±1.78	5.67±1.99
	0.05% P	2.80±1.15	3.07±1.03
	0.05%YE	3.73±1.22	4.60±1.12
	20% CW	1.27±0.46	1.73±0.59

Orchids are grown prominently as ornamentals and are valued as cut flowers not only because of their exotic beauty but also for their long shelf life. The propagation of orchids through green pod culture is widely followed for multiplication of species as well as hybrid orchids. Several workers have reported the advantage of using immature seeds from green pods for propagation [7]. Asymbiotic germination of orchid seeds depends on the use of an appropriate nutrient formulation. Out of the four different media tried for the present investigation Mitra *et al.*, medium was found effective in seed germination and subsequent protocorm development. The efficiency of Mitra *et al.*, medium for the seed germination was also reported in rare and endangered orchids of Western Ghats [8 - 10].

The beneficial effect of complex organic additives such as CH, P, YE on orchid seed germination and subsequent growth of protocorm was reported by earlier workers [11, 12]. However, the higher rates of seed germination in CH containing media may provide the developing protocorm with an instant source of nitrogen as amino acid which is more easily absorbed than the inorganic forms otherwise present in the medium. Protocorm proliferation leading to the formation of daughter protocorm as observed in the present study has also been reported to orchids like *Bletia urbana* [13], *Vanda spatulata* [9], *Ipsea malabarica* [8]. The protocorm proliferation is due to the meristamatic action of the protocorm. The pronounced formation of protocorm cluster in media with CH, P, W, YE obviously contribute to enhanced propagation rate. The high percentage establishment (83%) of the seedling indicated the healthy nature of the plant. The present protocol provides an efficient *in vitro* multiplication method for *A. rigida* from green capsule.

ACKNOWLEDGEMENTS

The authors are thankful to the Professor and Head, Department of Botany, University of Kerala for the facility provided.

CONFLICT OF INTEREST

The authors confirm that this chapter contents have no conflict of interest.

REFERENCES

[1] Nayar MP. Endemic flora of Peninsular India and its significance. Bull. Bot. Surv. India 1982; 22: 12-23.

[2] Gopalan R, AN Henry. Endemic Plants of India. CAMP for the strict endemics of Agasthiyamalai Hills, Southern Western Ghats. Dehra Dun: Bishen Singh Mahendra Pal Singh 2000; pp.476.

[3] IUCN. IUCN guidelines for the prevention of biodiversity loss due to biological invasion, Species 1999; pp.28-42.

[4] Mitra GC, Prasad RN, Chowdhury AR. Inorganic salts and differentiation of protocorms in seed callus of an orchid and correlated changes in its free amino acid content. Indian Journal of Experimental Biology 1976; 14: 350-351.

[5] Vacin E, Went FW. Some pH changes in nutrient solutions. Bot. Gaz 1949; 110: 605-613.
[6] Wimber DE. Clonal multiplication of Cymbidium through tissue culture of the shoot meristem. Am. Orchid Soc. Bull 1963; 32: 105-107.
[7] De KK, Majumdar S, Sharma R, Sharma B. Green pod culture and rapid micropropagation of *Dendrobium chrysanthum* Wall.- a horticultural and medicinal orchid. Folia Hortic 2006; 18: 81-90.
[8] Gangaprasad AN, Decruse WS, Seeni S, Menon S. Micropropagation and restoration of the endangered Malabar daffodil orchid *Ipsea malabarica.* Lindleyana 1999; 14: 38-46.
[9] Decruse SW, Gangaprasad A, Seeni S, Menon SV. Micropropagation and ecorestoration of *Vanda spathulata*, an exquisite orchid. Plant Cell Tissue and Organ Culture 2003; 72: 199-202.
[10] Parab GV, Krishnan S. Rapid *in vitro* multiplication of orchids *Aerides maculosa* Lindl. and *Rhyncostylis retusa* (L.) Bl. From immature seeds. Indian J. Biotechnology 2012; 11: 288-294.
[11] Bopaiah AK, SM Jorapur. Studies on growth and development of *Cymbidium aloifolium* Sw. Seedling *in vitro.* Orchid Wealth of India. In: Biology,Conservation and Culture of Orchids Eds. S. P. Vij. Affiliated East-West Press, New Delhi 1986; pp.423-427.
[12] Pathak P, Vij SP, Mahant KC. Ovule culture in Goodyera biflora (Lindl.) HK.F. a study *in vitro.* J. Orchid Soc. India 1992; 6: 49-53.
[13] Rubluo A, Chavez V, Martinez A. *In vitro* seed germination and reintroduction of *Bletia urbana* (Orchidaceae) in its natural habitat. Lindleyana 1989; 4: 68-73.

Chapter 26: Biodiversity Conservation - Challenges for Future and Way Forward

CHAPTER 26

Biodiversity Conservation - Challenges for Future and Way Forward

Preetha N., Laladhas K. P. and Oommen V. Oommen*

Kerala State Biodiversity Board, L-14, Jai Nagar, Medical College P.O., Thiruvananthapuram - 695 011, Kerala, India

Abstract: Although Protected areas (PA) are effective in Biodiversity governance a noticeable change to a landscape oriented approach and a '*People inclusive*' approach is observed. PA's are increasingly becoming isolated islands of wilderness areas in a matrix of domesticated landscapes and species. This has led to newer paradigms for conservation practices at the landscape level such as ecologically sensitive areas, community conserved areas. Adoption of new adaptive approaches may offer ways to further expand the protected area network since they allow for sustainable use of natural resources with people's participation. The economic potential of Biodiversity is unidentified which is considered as one of the factors for the rapid depletion of biodiversity and extinction of species. Financing biodiversity is vital for its conservation and management and mobilizing adequate finance is acknowledged as a huge challenge. Access and Benefit Sharing (ABS) is an emerging internal financial option. Successful models of conservation practices should be communicated, scaled up and replicated and a repertoire of pioneering models that can be employed in different circumstances should be developed. Indian experience of balancing conservation and development by employing a range of stratagem for biodiversity management is detailed.

Keywords: Benefit sharing, biodiversity governance, biodiversity loss, community reserves, corridor, ecologically sensitive areas, elephant, flagship species, fragmentation, hotspots, management, national park, protected area, tiger, tourism, wildlife sanctuary.

1. INTRODUCTION

"India's varied bio-physical environment, has led to the development of a rich heritage of bio resources.. India is home to about 8% of the recorded species of the world [1]. The pressure exerted on land due to population explosion, has in turn created unsustainable demands on bio resources [2]. Habitat loss, land use change and fragmentation are the major drivers of biodiversity loss globally.

***Corresponding author Oommen V. Oommen:** Kerala State Biodiversity Board, L-14, Jai Nagar, Medical College P.O., Thiruvananthapuram - 695 011, Kerala, India; Tel/Fax: 0471 - 2448234 (Off.); Mob: 9447728940; E-mail: oommenvo@gmail.com

Laladhas K.P., Oommen V. Oommen and Sudhakaran P.R. (Eds.)

Increasing pressure to meet the ever increasing demands of a growing population and a high growth trajectory is leading to immense land use changes in India. Overexploitation of natural resources mainly timber, medicinal plants, and other minor forest produce is a serious threat to biodiversity. Mining and quarrying has affected the fragile ecosystem of biodiversity hotspot areas like Western Ghats [3]. This chapter captures the broad challenges encountered during Biodiversity conservation and the way forward in the Indian context.

2. CONSERVATION PARADIGMS- CHALLENGES

2.1. Protected Area- Reservoirs of Biodiversity

The Conservation International has recognized 34 hotspots. Conservation efforts has yielded great success and has resulted in preventing at least 16 bird extinctions [4]. "India harbours four of the 34 global biodiversity hotspots Himalayas, Indo-Burma region, Western Ghats- Sri Lanka and Sundaland" as reported in India's 5th National report to the Convention on Biological Diversity, MoEF, Govt of India.

India has put into place a robust legislative and policy governance framework designed to conserve biodiversity. It has also launched major species conservation projects such as Project Tiger and the export and import of wildlife and their derivatives has been regulated through the umbrella Wild Life (Protection) Act, 1972. Protected areas (PA) occupy around 4.9% of the country's area presently and the area has expanded significantly in the past few decades under stringent management regimes. "The area covered under protected area network has increased from 155475.63 km^2 to 167408.3 km^2 during the period 2000 to 2014 [5] through an impressive figure of 102 National parks, 527 wildlife sanctuaries and 18 biosphere reserves".

2.1.1. Conservation Ethos in PA: Case Study

Traditional approaches to conservation have evolved over time and India Eco development Project was initiated during June, 1994 to conserve bio- diversity through eco development and aims at Improved Protected Area management [6]. The community based protection-oriented ecotourism program implemented at Periyar Tiger Reserve has promoted inclusive and participatory management and the programme has succeeded in reforming many poachers. The EDC have generated revenue exceeding INR 10.5 million from about 600,000 tourists annually for supporting Park management and also provides supplementary livelihood options to the dependent community [7].

Another key replicable biodiversity governance model in protected area management is Parambikulam Tiger reserve in Kerala where due to various joint activities of the park management and eco-development committee members' revenue amounting to Rs. 2,62,000 was generated during 2012 to 2013. Community based tourism, value addition of non-timber forest produce, organic certification and premium sale of agricultural products are some of the key community welfare initiatives undertaken by the park management [8].

Such innovative non-consumptive use of biodiversity in PA provides models for scaling up management practices in Protected area network nationally. For the perpetuity of PA's, management practice should evolve towards greater participation of community and grass root organizations for monitoring and evaluation at the same time ensuring benefit accrual mechanism through non-extractive use of biodiversity.

2.2. Flagship Conservation Programmes

India has special schemes for conservation of vulnerable ecosystems. Under the National Wetlands Conservation Programme, 115 wetlands have been identified for conservation of which 25 are acknowledged as Ramsar wetland sites. Twenty four marine protected areas have been established. An extent of 11,000 hectares is being conserved as Medicinal Plant Conservation Areas (MPCAs) in five states. Flagship species such as Tiger and Elephant are protected through 45 Tiger Reserves and 32 Elephant Reserves in the country [5]. In addition to tiger and elephant, recovery programmes for 15 other critically endangered species have also been launched recently. These include Asiatic lion, One-horned rhinoceros, Snow leopard, Asian wild buffalo, Malabar civet (*Viverra civettina*), Edible nest swiftlet *(Aerodramus fuciphagus)*, Hangul (*Cervus eldi eldi*), Nilgiri tahr, Manipur brow-antlered deer *(Rucervus eldii eldii)* and Swamp deer.

2.3. Ecologically Sensitive Areas

PA's are increasingly becoming isolated islands of wilderness areas in a matrix of domesticated landscapes and species. This has led to newer paradigms for conservation practices at the landscape level such as ecologically sensitive areas, community conserved areas. Four Community reserves, and 57 conservation reserves [5] have been established in India on an experimental basis. Ecologically Sensitive Areas (ESAs) have been identified and notified by MoEF since 1989 under the Environment (Protection) Act 1986 of India. ESA provides a people inclusive management regime based on sustainable development and equitable sharing of benefits for those areas vulnerable to ecological degradation. The decentralized governance model proposed by ESA opens up possibilities for making biodiversity management more responsive to ecosystem conservation with people's participation. Such a decentralized governance model has been proposed for Western Ghats where people have been considered as an integral part of the ecosystem [3, 9].

2.4. Biodiversity Heritage Sites

Taking into consideration the importance of local conservation initiatives and community participation, Biodiversity Boards are empowered to designate ecologically important areas as BHS as per The Biological Diversity Act 2002, Section 37. The State Biodiversity Board (SBB) in consultation with the local bodies, Biodiversity Management Committees (BMC) and other key stakeholders can recommend BHS without imposing any restrictions on the land use by locals.

3. BIODIVERSITY CONSERVATION – WAY FOREWORD

3.1. CEPA (Communication, Education and Public Awareness)

A major challenge to conservation initiatives is that with the exception of a few charismatic species, biodiversity issues are not relevant to public or policy makers. Biodiversity information is seldom used by local self government in planning process [10]. The Biological Diversity Act 2002 and Rules 2004 of India is a landmark legislation which regulates access to bio-resources, conservation of genetic resources and equitable sharing of benefits arising out of it. The Act is being implemented at three levels, National Biodiversity Authority (NBA), State Biodiversity Board (SBB) and Biodiversity Management Committees (BMC). The establishment of BMC at local level puts in place an institutional mechanism by which the Biological diversity data generated through People's Biodiversity Register is integrated into biodiversity conservation concerns for local level planning. Mainstreaming biodiversity issues at local level planning and prioritizing the areas/species for conservation and commercial utilization will result in greater focus on conservation and resource management for sustainable development.

3.2. Biodiversity Governance

Protected areas and community based conservation are the two major streams of governance models attempted globally for biodiversity conservation. "Although 12% of Earth's land surface has protected-area status, only 0.5% of oceans and

5.9% of territorial seas have been so designated, and more than two-thirds of critical sites for biodiversity have incomplete protection or none at all" [11]. Community conservation areas are new models of biodiversity governance [12]. "Protected areas need to be managed as a coherent network rather than as isolated habitat islands in order to sustain biodiversity" [13]. Main barriers to effective protected area governance includes: (i) land-use issues such as limited land per capita; (ii) human issues and loss of livelihood; (iii) institutional issues; (iv) Man-animal conflicts. Biodiversity conservation is the backbone of protection of inviolate forest area and strict guidelines for identification of such areas are necessary. Biodiversity rich areas outside protected areas should be identified and conserved to combat forest fragmentation as they can act as corridors, connecting wildlife habitats.

3.3. Forests Fragmentation and Biological Corridors

In densely populated countries such as India forest fragmentation is the key factor for loss of ecosystem continuity. Fragmentation isolates and reduces the size of populations and in many cases territorial species such as Tigers in search of territory migrate to the fringes of forest, encroaching upon populated place. PA's are either too small to safeguard viable wildlife populations or act as islands in a sea of human dominated land uses. Biological corridors can increase the functional landscape connectivity among fragments and decreases the extinction rate by increasing inter fragment movements. Such corridors can be identified for the most extinction prone species based on ecological information and seasonal movements of the target species at the same time taking into consideration the impact on human activities and land uses.

3.4. Flagship Species

Although Biodiversity conservation encompass the variety of living organism present in the world, priority actions to conserve *flagship species* are important.

3.5. Development of a National Biodiversity Grid (NBG)

Biodiversity related information is available in India in more than 100 databases and networks in various forms such as databases, GIS maps, specimens, herbaria *etc*. The Grid, that will be established as a part of the National Knowledge Network (NKN) with assistance from Centre for Development of Advanced Computing (C-DAC) will serve as a unique technology based platform for linking biodiversity based data and information across the country. The NBG project was mooted in the draft National Biodiversity Information outlook released during CoP11 as a joint effort by MoEF and NBA to address the conservation challenges in India and contribute to economic growth, ecological sustainability and social well being. NBG will assist in policy making through analyzing nation-wide biodiversity data.

3.6. Valuation of Biodiversity

The economic potential of Biodiversity is unidentified which is considered as one of the factors for the rapid depletion or degradation of biodiversity and extinction of species. Gross Domestic Product (GDP) takes into account NTFPs and timber from forestry but the various ecosystem services provided by forests are not taken into consideration.

Most of the biodiversity values are implicit rather than explicit. In ABS perspective, measuring the value of bio-resources/products used for manufacturing different production process is urgently needed. Proper valuation of ecosystem services is necessary [14].

3.7. Financing Biodiversity

Financing biodiversity is vital for its conservation and management and mobilizing adequate finance is acknowledged as a huge challenge. Access and Benefit Sharing (ABS) is an emerging internal financial option.

3.8. Access and Benefit Sharing

The challenge with Access and Benefit sharing agreements is to identify and ensure that the benefit accrual mechanism trickles down to the rightful local "benefit claimers"- individuals or community. Several State Biodiversity Boards have taken the initiative to empower BMC's to manage their natural resources sustainably. For instance, in Madhya Pradesh, SBB facilitated an ABS agreement for kalmegh (*Andrographis paniculata*) between the Malajkhand municipal BMC in Balaghat and Natural Remedies (manufacturers of ayurvedic veterinary products and herbal medicines). As a result, 10 tonnes of kalmegh had been collected and sold at Rs. 25 per kilogram as of March 2012. In Andhra Pradesh, Bio-India Biologicals (BIB), Hyderabad, paid Rs 100 per kg for transport of neem leaves to Japan from the Amarchintha village BMC in Mahabubnagar district and the Revalli BMC in Nalgonda to produce neem based water [7].

3.9. Traditional Ecological Knowledge

Farmers' traditional knowledge is the key to sustain biodiversity by maintaining agrodiversity of crop wild relatives to deal with changes of soil, climate, nutrition, flavour, storage conditions, pests and diseases. Managing and enhancing diversity of landscape through TEK is insurance for resilience, the ability of ecosystem to recover from changes to continue to function and provide ecological services. In order to build synergies among knowledge systems for the benefit of biodiversity conservation, indigenous knowledge will help in drawing strategies for the management of biodiversity and ecosystem services [15].

4. LEARNING FROM LESSONS

Many species in nature are becoming isolated from each other; conservation biologists are identifying critical co evolutionary links, between keystone species essential for ecosystem stability. Targeted conservation measures for such keystone species have to be undertaken on a priority basis. Conservation programmes ahould involve the local people [16]. India's strength in biodiversity governance lies in its varied approaches both in terms of PA centric and landscape centric approaches. Protected areas continue to be effective tool for conservation. Adoption of new adaptive approaches such as Conservation Reserves, ESA, BHS *etc* may offer ways to further expand the protected area network since they allow sustainable use of natural resources with peoples participation. Neither Protected areas nor community-based conservation is a panacea for all conservation problems and community conserved areas could be considered under both protected areas and community-based conservation [17]. Successful models of conservation practices should be communicated, scaled up and replicated and a repertoire of pioneering models that can be employed in different circumstances should be developed. In this context, the age old and varied Indian experience of balancing conservation and development by employing a range of stratagem for biodiversity management has immense international relevance.

ACKNOWLEDGEMENTS

Declared none.

CONFLICT OF INTEREST

The authors confirm that this chapter contents have no conflict of interest.

REFERENCES

[1] India's Fourth National Report to the Convention on Biological Diversity. Ministry of Environment and Forests, Govt of India, New Delhi 2009.

[2] Laladhas KP, Preetha N, Baijulal B, Oommen V Oommen. Strategies for Biodiversity Governance – Kerala Perspectives. International Conference on Environmental Biology and Ecological Modelling (ICEBEM 2014), Santiniketan, Programme Abstracts 2014: pp. 28.

[3] K Kasturirangan Report of the High Level Working Group on Western Ghats Ministry of Environment and Forests, Government of India, 15 April 2013.

[4] Butchart HM, Stattersfield AJ, Collar NJ. How many Bird extinctions have we prevented Oryx 2006; 40: 266.

[5] www.wiienvis.nic.in.

[6] Protected Area Management in India: A Perspective Maitreyi Mandal XII World Forestry Congress at Quebec, Canada, September 21-28, 2003.

[7] Seema Bhattt, Kabir Bavikattee, Suneetha M Subramanian. Comuunity based experiences on access and benefit sharing case studies National Biodiversity Authority 2012.

[8] UNDP –India 2014, India Biodiversity Awards, 2014.

[9] Madhav Gadgil. Report of the Western Ghats Ecology Expert Panel, Ministry of Environment and Forests, Government of India, 31 August 2011.

[10] Cort CA. A survey of the use of natural heritage data in local land-use planning. Conservation Biology 1996; 10: 632– 637.

[11] Butchart SHM *et al.* Science 328, 1164 (2010); Global Biodiversity indicaors of recent declines. published online 29 April 2010 (10.1126/science.1187512).

[12] Brechin SR, Wilshusen PR, Fortwangler CL, West PC. Promoting International Biodiversity with Social Justice in the Twenty-first Century (State Univ. of New York Press, Albany, 2003).

[13] Hole DG, *Willis SG, Pain DJ* et al. *Projected impacts of climate change on a continent-wide protected area network.* Ecology Letters *2009*; 12*:* 420-*431.*

[14] Prakash Nelliyat, Balakrishna Pisupati. Valuation of Biodiversity. National Biodiversity Authority 2013.

[15] Laladhas KP, Preetha N, Baijulal B, Oommen V Oommen. Building Synergies Between Science and Local Knowledge for Conservation of Biodiversity – People's Biodiversity Register, Compendium – Traditional Knowledge 26th Kerala Science Congress, Kerala. pp. 122-142. 2.

[16] Schaller GB. The Last Panda. University of Chicago Press, Chicago and London1993. pp. 35.

[17] Pramod Krishnan, Rajesh Ramakrishnan, Sushil Saigal, Shailesh Nagar, Faizi S, Panwar HS, Subrata Singh, Nimesh Ved. Conservation across Landscapes, India approaches to Biodiversity governance UNDP 2012.

Authors Index

A

Anaswara Krishnan S.
Anilkumar A.
Anilkumar C.
Anisha Shashidharan
Anjusha S.
Ashok Verma

B

Baijulal B.
Balaram P.
Bharadwaz M.
Bindu S.
Blessan Santhosh George

C

Chitra C. R.

D

Damodaran A.
Daniels A. E. D.
David J. Gower
Debal Ray
Debbarma J.
Devanesan S.
Dey T.
Dileepkumar R.
Dutta P.

G

Ganesan M.
Gangaprasad A.
Geetha Antony
Geetha P. M.
Gimy George
Gogoi K.
Gupta A. K.

I

Islam S.J.

J

Jacob J.
Jasmine S.
Jayamani C. V.
Joshi K. K.

K

Kalita J.
Kalita M.
Kariyappa K. C.
Krishnakumar N.
Kumar Awadhesh

L

Laladhas K.P.
Lekha Divya

M

Manohara T. N.
Mark Wilkinson
Mary P. P.
Mathish Nambiar-Veetil
Maya P. V.
Mohammad A. Akbarsha
Mohammed-Aslam M. A.
Mohankumar Chinnamma
Molly Varghese

N

Narayanakumar R.

O

Oommen V. Oommen

P

Prajith T. M.
Preetha N.
Premila K. S.

R

Raghunath T. P.
Rajasekharan S.
Rajmohan K.
Ramachandran Kotharambath
Ranimary George
Razia Beevi M.
Renjan Mathew Varghese
Ronald A. Nussbaum

S

Salini Bhasker
Sanil George
Sanila V. K.

Laladhas K.P., Oommen V. Oommen and Sudhakaran P.R. (Eds.)

Saravanan S.
Selvaraj P.
Shailaja K. K.
Shereif A. K.
Sibin N. P.
Smitha K. P.
Somy Kuriakose
Soumyendra Nath Ghosh
Sreejith Aravindakshan
Srinath K. R.
Sudarslal S.
Sudipta Mukherjee

U

Unni B.G.

V

Vandana Thaplyal

Subject Index

A

Acampe rigida 241
Access and benefit sharing (ABS) 60, 249, 254
Ad valorem rates 169
Agricultural income tax 160, 165, 167
Agricultural practices 127-28
Agriculture
  high external input 137-38
  specialized 147
Airsheds 166
Alappuzha 190, 196
Alcohol 169-70
Alkylation 210-11
Allergy 78, 81
Allium cepa 223-24
Alpha diversity 65
Amino acid 103, 106, 213, 246
Amphibians 41-43
  species of 41, 43
Amphibian species 41-43
  threatened 43
Amphizoa sp 237-38
Anacardiaceae 115-16, 223-24
Anaimalai earth snake 8
Ancient 53-54
Ancient Indian concepts 54
Ancorinidae 186
Animals, stray 200-201
Antheraea assama 81-82
Anthurium 225-26
Antibacterial activity 71, 85
Antimicrobial activity 71
Antitumor 103-4
Aquatic sow bug 237-38
Arachnida 235-37, 240
Araena sp 236-38
Archilejeunea apiculifolia 232
Areas
  biodiversity hotspot 250
  natural 89-90, 166-67, 170
  Arginine 104-5
Aromatic plants 28, 56-57
Assam finance tax 164-65
Associated traditional knowledge 53, 152-53
Asthma 77-80, 200, 225
Averrhoa bilimbi 223-24
Awareness programmes, special 101
Ayurvedic arboreal medicine 54
Ayurvedic industry 27-28
Ayurvedic veterinary medicine 54
Azadirachta indica 55, 222-23

B

Balancing conservation 249, 255
Bamboo shoots 121, 123, 125
Bark of trees 229-30
Beach erosion 195, 199
Beaches, nesting 197, 199
Beach sand mining 195, 199
Beddome 6-8, 10, 13
Beetles 138, 143-44
Beneficiary farmers 137, 139
Benefits of biodiversity conservation 65, 158
Beta diversity 65
Binding pockets 103-4
Bio-control agents 127, 130
Biodiversity
  agricultural 66
  conserve frog 41
  conserving agro 172
  dry land 66
  financing 249, 254
  forest 66
  global 23
  inland water 66
  linking 253
  marine 66
  mountain 66
  penalize 158
  plant 28
  regional 91
  rich 113, 221
  safeguard 67
  sustain 253, 255
  terrestrial 54
Bio-diversity conservation 5, 31, 65
Biodiversity Conservation financing 158
Biodiversity Conservation of modern
  concepts/strategies of 53-54
Biodiversity conservation bias 163
Biodiversity conservation goals 158
Biodiversity conservation initiatives 69
Biodiversity conservation instrument 163, 168
Biodiversity conservation programmes 55, 67
Biodiversity conservation strategies 54, 67
Biodiversity conservation targets 67
Biodiversity conservation ventures 167
Biodiversity databases 48
Biodiversity enrichment 137-39
Biodiversity focus 158

Laladhas K.P., Oommen V. Oommen and Sudhakaran P.R. (Eds.)

Biodiversity fund 35, 160, 167
Biodiversity governance 249, 252-53, 255
Biodiversity heritage sites 252
Biodiversity importance 166
Biodiversity indicators 183, 189
Biodiversity information 48, 184, 252
Biodiversity issues 252
Biodiversity legislations 27
enforcement of 27, 38
Biodiversity management 53, 249, 252, 255
Biodiversity Management Committee (BMC) 27, 32-36, 38, 152-53, 252, 254
Biodiversity management committees 27, 152, 252
Biodiversity offsets 172
Biodiversity of north eastern region 72
Biodiversity outcomes 171
securing positive 157
Biodiversity protection 165
Biodiversity registers 27
Biodiversity reserves 160
Biodiversity resources 157
Biodiversity use 167
Biodiversity values 69, 178, 191, 254
Biodiversity variation 139
Bio-India Biologicals (BIB) 254
Biological diversity 32, 71, 151, 159, 171, 191, 250
Biological diversity act 32, 61, 151-52, 156, 252
Biological resources 32, 35, 68, 84, 157, 159, 170, 184
Bioresources of north eastern region 83-84
Biosphere reserve 56, 67, 250
Biotechnological tools 68
BLf gene 104-5
BMC for resource management 27, 38
Bombyx mori 73, 82
Bos indicus 103-4
Bos Taurus 103
Botanical garden 57, 71, 83
Bovine lactoferrin (BLf) 103-5
Bronchitis 79-80, 85
Bryophyte diversity 230-31
Bryophytes 229-31, 233
Burmese coriander 222, 224

C

Calamus 114, 117, 121-23, 125
Campanulata 229, 232-33
Cannon ball trees 226
Capped langur 75, 97-98
Carboxamidomethylated peptide fractions 211
Carboxamidomethylation 209, 211
Caridina typus 237-38, 240
Casuarina Plantations 195, 199
Cattle egrets 145-46
cDNA library 65
Cedrus deodara 93, 95
Central Board 162
Charges for Biodiversity Conservation 163
Chrysopelea ornata 9-10
Clitoria ternatea 209
Coastal afforestation 195, 201
Coastal areas 168, 195-97
Coastal biodiversity 195
Coastal development 195, 200
Coastal districts 195-96, 200
Coastal pollution 195
Coconut water (CW) 241, 243-45
Coding region of Lf gene 104
Coelentrates 184-85
Coffea arabica 222-23
Coffea robusta 222-23
Coleoptera 145, 204
Community participation 55, 121, 125, 219, 251-52
Community reserves 249, 251
Computer-assisted learning (CAL) 47
Conservation
agrodiversity 156
community-based 255
ecosystem 84, 252
environmental 160
ex-situ 54, 68
livestock genetic resource 65
snake 24
turtle 195, 198-99
Conservation activities 101, 152
Conservation area networks (CANs) 23
Conservation biologists 23, 255
Conservation efforts 167, 250
Conservation groups, local 195, 198, 200
Conservation lands 166, 172
Conservation measures 23, 97, 186, 255
Conservation of biological resources 159
Conservation of bio-resources 151
Conservation planning 3, 19
Conservation practices 54, 249, 251, 255
Conservation reserves 251, 255
Conservation strategies 3, 54, 68, 83, 97, 101
effective biodiversity 67
Conserve biodiversity 57, 65, 69, 91, 250
Conserve flagship species 253
Conserve Indian gaur 97, 101
Consolidated fund of India 162
Constitution of India 45
Convention of Biodiversity (CBD) 59-60, 151-52, 184
Coral reefs 163, 185

Correlation of species distribution 114
Corticolous 229, 232
Cottage industries 164
Crop rotation 127
Crops 66, 69, 82, 152, 221
  horticultural 113, 129
  vegetable 133, 222-24
Crustaceans 184-86
Cryopreservation 65, 68
Cultivation
  rattan 121, 125
  shifting 121, 125
Culture
  frog 47-48
  green pod 241, 245
  tissue 65, 68, 241-42
Cycles, input-output 137-38
Cyclic peptide 209-11

D

Daboia russelii 4, 13, 18
Damsel fly- *nymph* 238
Data, nation-wide biodiversity 254
Data envelopment analysis 131-32
Decentralized governance models 252
Decision management units (DMUs) 132
De novo sequencing 209
Dental problem 79-80
Development paradigm 53, 138
Diabetes 78-79, 81, 85
Diarrhea 77-80
Dipterocarpaceae 117
Direct tax 165
Disease resistance 103, 110
Disease resistant hybrids 86
DNA library 53
Dominant plant species 91
Dragon fly- *nymph* 237-38

E

Echis carinatus 13, 18
Eco development committees (EDCs) 195, 201, 250
Eco-friendly cultivation 137, 143
Ecological adaptations 113, 118
Ecological balance 178
Ecological conditions 71, 84, 127, 137
Ecologically Sensitive Areas (ESA) 249, 251-52, 255
Ecological status 3, 5, 7, 9, 11, 13, 15, 17, 19, 21, 23
Economic growth 60, 177-79, 254
Economic impact 131, 133
Economic incentives 31, 157, 160-61
Economic instruments 157-58, 160-62, 168, 170, 173
Economics, welfare 176-77
Economics of ecosystems and biodiversity 173
Economics resource 176
Economic valuation 184
Economic values 184
Ecosystem balance 137-38
Ecosystem based fisheries management (EBFM) 191-92
Ecosystems 3, 23, 31, 65-67, 69, 84-85, 90, 95, 147, 158-59, 163, 177, 191, 252, 255
Ecosystem services 53, 66, 173, 183, 191, 254-55
Edible rattans 121-22, 125
Edible shoots 121
Eichhornia 236-37, 240
Eichhornia crassipes 235-38, 240
Eichornia 146-47
Electrospray 209, 211
Elettaria cardamomi 222, 224
Embryo culture 241
Endemic animals and birds of North-East India 74
Endemic animal species 75-76
Endemic plants 71-72, 74, 83, 89
Endemic plant species 74
Environmental behaviour 180
Environmental economics 175-78
Environmental taxes 165
Epilepsy 77-79
Eri silk worms 73
Euphlyctis cyanophlyctis 44, 46-47
Euphlyctis hexadactylus 44, 46-47
Euphorbiaceae 223, 225-26
European patent office (EPO) 59
Evergreen forests 98, 229-30
Excise duties 165, 168-69
Excise duties biodiversity friendly 169
Excise tax 168-69
Exclusive economic zone (EEZ) 191
Experience, varied Indian 255
Exploitation 3, 101, 121, 175, 178, 190
  economic 83
Ex situ conservation 57, 68, 83-84

F

Fair trade certification in India 32
Fauna 71, 74, 85, 90, 137-40, 143, 179, 185, 188, 235-37
  endemic species 75
Faunal biodiversity variation 139
Federation of Chambers of Commerce and Industry (FICCI) 122

Fees, seigniorage 163
Fenneropenaeus indicus 236-37, 240
Fenneropenaeus indicus macrobrachium 238
Fertilizers 42, 133, 138, 144
Fibres, dietary 121-22
Ficus benghalensis 55, 95
Financial resources 172
Financing mechanisms 157
Fiscal instruments 157, 172
Fishery resource assessment division (FRAD) 184
Fishing harbours 195, 200
Fish species 183, 187
Flagellum 122-23
Flora and fauna 74, 85, 137-40, 143, 179, 188
Floral species diversity 183
Flowering plants 72, 74, 222, 242
Food and agriculture organization (FAO) 84, 104
Food plants 57, 76
Food quality 127-28
Food safety 127-29, 133
Food safety regulations 128
Food security 123, 125, 137-38, 226
Forest ecosystems 57, 69
Forest landscape 203-4
Forest patches 67, 90-91
Forest produce 164, 166
Forest resources 121, 164-65
Forest royalties 165
Forests 19, 21, 36, 45, 56, 71-72, 89-90, 94-95, 125, 160, 163-66, 170-71, 173, 179, 186, 201, 204, 206, 241-42, 253-54
  bushy 97, 99-100
  conservation of 90, 166
  degraded 97, 99-100, 232
  dense 97, 99-100
  minor 164, 250
  natural 167, 169, 204
Form
  corticolous 232
  terricolous 232
Fossombronia himalayensis 232-33
Frogs 4, 41-48, 144-45, 206
  farmed 48
  green 44, 46-47
Frog species 44, 46
  breed Indian 46
  large edible 44
Fruit crops 222-24
Frullania acutiloba 229, 232-33

G

Gamma diversity 65
Garcinia gummi-gutta 113, 117-19
Garcinia gummi-gutta dormancy 118
Gastropoda 235-38
Gene banks 53, 65, 67-68
Gene pool conservation area (GPCA) 57
General agreement on tariffs and trade (GATT) 152
Genetic diversity 65, 67-68
Genetic resources 68, 86, 151, 159, 252
  aromatic plant 57
Genetic stock 29-30
Global amphibian assessment (GAA) 43
Global biodiversity information network 65
Global positioning system (GPS) 89, 91
Glycoprotein 103-4
Gmelin 9
GMP certification 27, 32, 35, 38
Golden cat 74-75
Good manufacturing practices (GMP) 27, 32, 35
Goods 84, 154, 160-62, 168, 170, 176, 180, 191
  biodiversity-related 173
  environmental 179
  non-merit 169-70
  sumptuary 168-69
Gorgonids 184, 187
Gout 77-78
Green turtle 195, 197
Gross domestic Product (GDP) 254
Gynecological disorders 78-79

H

Habitat banking 172-73
Habitat degradation 41, 43
Habitat destruction 3, 19-20, 30, 121, 240
Habitat displacement 45-46
Habitat diversity 28, 232-33
Habitat loss 19-20, 27-28, 42, 65, 67, 84, 190, 195, 199, 201, 249
Habitat management 24
Habitats, based 172
Haemadipsa cochiniana 236-38
Harvesting
  destructive 27-28, 30
  management of 33-34
Harvesting permit 34
Hevea brasiliensis 222-23
Hexadactylus 46, 48
High performance liquid chromatography (HPLC) 209-10, 212
Himalayan Goral 75-76
Himalayas, global biodiversity hotspots 250
Hirudinaria granulosa 237
Hog Badgers 75
Horticultural products, export of 129
Host plants 82-83, 86

Hot spots of biodiversity 66
Human Capacity to Conserve Biodiversity 65
Hydrilla 145-46
Hygrophila 145-46
Hyophila comosa 229, 232-33
Hypertension 78, 80
Hypnale hypnale 4, 13, 18

I

Income tax 160, 164-65
India biodiversity reserves 160
India Eco development Project 250
India forest fragmentation 253
India forest management 83
India harbours 250
Indian and Thai rattans 125
Indian bison 97, 99, 101
Indian black earth snake 6
Indian civet 76
Indian cobra 12
Indian earth snake 7
Indian elephant 75
Indian fisheries Act 152
Indian forest act 160
Indian gamma snake 11
Indian gaur 97-98, 100
Indian green keelback 10
Indian ornate flying snake 9
Indian plum *Ziziphus* 77
Indian pond frog 44
Indian Python 8
Indian rat snake 8
Indian rock python 8
Indian skipper frog 44
Indian slender coral snake 12
Indian snakes 24
Indian state of Maharashtra 129
Indian students of biology 44
Indian subcontinent 3-4
Indian system 28
Indian system of medicine 32
Indian vulture 75
Indian wart snake 8
Indian wildlife 97-98
India ranks twelfth 66
India's coastal zone regulations 168
India's decision 160
India's medicinal plants 30
India's planning process 171
India's strength 255
India's water cess act 162
India's wetlands 163
Indigenous knowledge 57, 97, 101, 255
Indigenous seeds 113
Indigofera tinctoria 222-23
Indirect tax 157-58, 165, 168
in situ 53-54, 67
Inspite of India's status 129
Instruments, biodiversity conserving 161
Integrated pest management (IPM) 128, 130
Intensive farming 137-38
International year of biodiversity 69
Invasive species 69
  native 30
Isopterygium lignicola 232-33
Ixora coccinea 222-23

K

Kalimpongensis 229, 232-33
Kasaragod, coastal regions of 217, 219
Kasaragod district 218
Kerala burrowing snake 7
Kerala on marine turtles 197
Kerala state biodiversity board and national biodiversity authority 38
Kerala state federation of scheduled caste and scheduled tribes development co-operatives limited 31
King cobra 4, 12, 20

L

Lagerstroemia speciosa 205
Lamellidens marginalis 237
Lampyridae 204
Large Indian Parakeet 76
Leopard cat 75
Leucoderma 78-79
Lf gene 104
Lf protein 104
Light tolerant 229-30, 232-33
Light tolerant species 229
Live green 27, 35, 38
Local level planning 252
Lowering cholesterol 121-22
Lowsonia alba 223
LSG level species wise/habitat wise 34
Luffa cylindrica 223-24
Lumbago 79-80

M

Macrobrachium sp. 236-37, 240
Macro invertebrates 235-39
Macrophytes 235-36

Mainstreaming biodiversity issues 252
Management practices 251
  integrated pest 130
Mangifera indica 55-56, 223-24
Mangroves 163, 184, 188-89, 205, 217-18
  conservation of 217, 219
  destruction of 188
Manipulation zones 67
MAP resources 29, 33-34, 36, 38
MAP species 29, 31, 38
Marbled cat 75
Marigold 225-26
Marine mammals 184, 189-90
Marine turtles 195, 197, 199, 201
Market, domestic 28, 129
Mass extinctions 41-43, 66
Mass spectrometric biodiversity conservation 211
Mastigolejeunea auriculata 232
Mastitis 105
Maximize biodiversity representation 23
Measuring biodiversity 183
Medicinal and aromatic plants (MAP) 27-31, 33-34, 36, 38
Medicinal plant conservation areas (MPCAs) 251
Medicinal plant parts 30, 34
Medicinal plants 27-32, 35, 71-74, 76-77, 85-86, 89-90, 221-23, 250
  common 76
  species of 28, 72-73
Medicinal plants board 33, 36
Medicinal plants trading society 27, 33, 38
Medicinal properties 85, 103, 221
Medicinal species 28
Mega-biodiversity countries 41, 43
Meliponiculture 221, 226
Microbial communities 84, 86
Microcionidae 186
Micropropagation 241
Mimosa pudica 79, 223
Models of conservation practices 249, 255
Mongooses 75-76, 199
Monochoria 236-37, 240
Monochoria vaginalis 235-38, 240
Mrigayurveda 54
Muga silkworms 76, 81-82
Mulberry 73, 81-83
Mussel
  brown 187
  green 187

N

National biodiversity act 160
National biodiversity action plan (NBAP) 48, 160, 163
National biodiversity authority (NBA) 32, 59, 153, 252, 254
National biodiversity fund 158, 171
National biodiversity grid (NBG) 253-54
National biodiversity legislation of India 171
National knowledge network (NKN) 253
Natural biological control plot (NBCP) 138, 143
Natural habitat 20-21, 24, 57, 68-69, 84, 242
Natural resources 28, 30, 36, 38, 61, 74, 84-86, 89-90, 95, 101, 147, 162, 172, 177-79, 250, 254
  conservation of 175, 178
  sustainable use of 249, 255
Nature conservation 90
Niacin 121-22
Non-dedicated tax 165-66
Non-venomous 7, 9, 11
Non-venomous data 6-10
Non-venomous rare 6, 8-10
Northeast Indian states 98
North East region of India 85
Nucleotide polymorphism 103-4
Nutritional care, special 54
Nymphaea 145-46

O

Ocimum sanctum 222-23
Odayam 197
Oil seeds 222, 224
Oligochaeta 236-37
Ophiophagus Hannah 4, 12, 17
Optimum utilization of biodiversity 67
Orchids 226, 241-42, 245-46
Ordinary least square (OLS) 131
Organic additives 241, 244-45
Organic farming 127-28, 130
Organisms
  marine 184-85
  species of 185, 236
Orthodox 115, 117
Over-exploitation 20, 27-28, 30
Overharvesting 41-42, 44
Overharvesting of frogs 44-45
Over the counter (OTC) 28

P

Penaeid prawn 183, 191
Peninsular India 46
Pennisetum typhoides 223-24
Peptides 209-13
  rBLf 109
Peptone 241, 244
Performance liquid chromatography, high 209-10

Pesticides 42, 69, 129-31, 133, 138, 144, 230
Pesticide safety 127-30, 133
Pest management 127, 129-30
integrated 130
PETA-India 46
Pheasants 74-75, 95
Philonotis fontana 232-33
Phyllanthus emblica 55
Phyllanthus niruri 222-23
Phytoplankton 184, 190
Pigouvian taxes 163, 165-66
Pigovian taxes 166
Pig-tailed macaque 75, 97-98
Pinus roxburghii 92, 95
Planned plantation activities 217, 219
Planorbis corneus 236-38
Plantation crops 222-23
Plant diversity 84, 209
Plant extracts 85, 210
Plant material 30
Plants 28, 30-31, 34, 41, 43, 53-58, 66, 68, 72-73, 76-79, 84-86, 90-91, 93, 95, 119, 146-47, 162, 209-10, 221-23, 226, 236-37, 240-42, 246
foraging 221, 226
forest 222
ornamental 222-23, 225
shade loving 231
Plant species, silkworms host 84
Plant tissue culture 242
Poisonous 4
Pokkali rice cultivation 190
Pollen bank 53
Polluting activities 195, 201
Polymerase chain reaction (PCR) 86, 104
Polythene bag 235
Popowia beddomeana 116-17
Population density 23, 34, 97-98, 100
Population density of Indian gaur 100
Porcupines 75, 95
Post monsoon 218, 235-37
Potential immunostimulant 103-4
Pox, small 79
Practices, chemical input-intensive yield-enhancement 127-28
Predacious diving beetle 237-40
Prepaid licenses 164
Price, specified 33
Problems 46-47, 69, 77, 115-16, 129, 146, 165, 172, 188
dominant 127-28
throat 77, 79-80
Production
crop 127-28
sustainable 176-77
Production-distribution chain 170
Products, agricultural 170, 251
Programmes, bio-diversity conservation 159
Propagation 31, 54, 245-46
Property, bactericidal 103-5
Property tax 160, 165, 167-68
higher incidence of 168
Protease 211-12
Protected area network 249-51, 255
Protected area- reservoirs of biodiversity 250
Protected areas (PA) 24, 56, 67, 155, 160, 168, 249-53, 255
Protection of TK 59-60
Proteins 121-22
crude 123-24
Protocorm 241, 244-46
daughter 244, 246
developing 241, 243, 246
Protocorm development 241, 246
Protocorm proliferation 244-46
Protozoans 190
Puncture vine 145-46

Q

Quality timber 165
Quilon 187-88

R

Random amplified polymorphic DNA (RAPD) 68, 86
Rates 28, 35, 42, 44, 65, 71-72, 162, 164, 167-69
non-merit 169
present species extinction 66
Rattan shoots 121-22, 125
Rattan species 121, 125
edible 122, 125
RBLf protein 103-4
Red List of Threatened species 98
Remote sensing 217, 219
Replicable biodiversity governance models 251
Research, agricultural 127-28
Residues, pesticide 127-28
Resource consumption 69
Resource economics 177
Resource management 27, 33, 35-36, 38, 252
Resource management agencies 33, 36
Resource mobilization 35, 157, 172
Respiratory diseases 77-79
threatened 217, 219
Respondents, perception of 138-39
Revenue generation 168

Reversed phase column 210
Rhesus macaque 75, 97-98
Rheumatism 78-80
Rice cultivation 138-39
Rice eco-system 137-39, 147
Rice farmers 146-47
Rice fields 137, 139, 143, 145-46
Rice varieties, traditional 151, 153, 156
Ricinus communis 222-23
Ridley, Olive 195, 197
Rubber plantation 20, 229-30, 232
Rubiaceae 223, 225
Rupicolous 229, 232

S

Sacred forest 65, 67
Sacred groves 53, 56, 89-95
Safety declination 127-28
Salamanders 42
Sales tax and value 170
Sales tax system in India 170
Salt tolerant ecosystem 217
Sector, vegetable 127-28
Seed bank 83, 113
Seed coat layers 118
Seed germination 241, 243-46
Seed samples 113
Seigniorage 157, 163-64
Sericulture 71, 73, 81, 83, 86
Sericulture biodiversity 83
Silk moths 73-74
Single-nucleotide polymorphism (SNPs) 68, 103-4, 106
In situ biodiversity 161
In situ conservation 56, 67, 95
Skin disease 79-81
Snake bite 77, 80
Snake population 20, 23
Snake species 3-6, 19-20, 23
  critical 23
  rare 20
  terrestrial 20
Special Issue of Eco-chronicle 219
Species 3-4, 20-21, 23-24, 28-32, 34-35, 41-44, 46-47, 53-54, 65-69, 71-75, 77, 84, 95, 97, 101, 113-14, 118, 121-22, 125, 158, 164, 183-89, 197, 206, 221, 229-32, 235-38, 240-42, 245, 249, 251, 255
  animal 3, 53, 74
  bamboo 72, 74, 121-22
  bee foraging 222
  carnivorous nocturnal 74
  charismatic 252
  cold blooded 4
  common Southeast Asian 205
  deciduous 230
  delicate bryophyte 233
  endemic 3, 48, 67, 74-75, 90
  endemic *Calamus* 114
  exclusive 230
  exotic 66
  extinction prone 253
  faunal 19
  flagship 249, 251, 253
  germinable recalcitrant 115
  harvested 31
  important mammal 76
  inhabiting 196
  insect *Antheraea* 82
  keystone 255
  mammalian 76
  molluscan 183, 187
  moss 232
  near-endemic 76
  recorded 53, 249
  reptilian 4
  rhododenderon 74
  sacred 95
  saving 68
  target 34, 253
  tree 54, 92
  un-harvested 165
  venomous 4
Species abundance 23, 236-37
Species *Bos indicus* 104
Species composition 203, 206
  unique plant 203
Species Corticolous 232
Species Corticolous Rupicolous Terricolous 233
Species distributions 23, 114
  snake 23
Species diversity 3, 19, 65, 68, 185, 191, 229-30, 237-38
Species-habitat associations 23
Species-habitat associationship 3
Species-habitat relationship 23
Species hinders 5
Species of birds 74, 76, 189
Species of bivalves 183, 187
Species of canes 72, 74
Species of cephalopods 183, 187
Species of crabs 183, 185, 187
Species of flowering plants 72, 74
Species of gastropods 183, 187
Species of gymnosperms 72, 74
Species of lobsters 183, 185, 187

Species of primates 74-75, 97-98
Species of scaphopods 183, 187
Species of shrimps 183, 185, 187
Species Perna viridis 187
Species regeneration protocols 118
Species restoration 119
Species Scomberomorus commerson 188
Species survivability 117
Specific-rate excise tax 169
Specimens
  gathered 31
  magnificent 91
  preserved 22
Specimen suppliers 46
Stages of economic history 178
State forest acts 163, 166
Stingless bee flora 221, 223, 225-26
Stingless bees 221-22
Stomach 78
Stomach disorders 77-80, 85
Strategic plan for biodiversity 172
Strategic plan on biodiversity conservation 157
Stratum-wise population density 97, 100
Suberitidae 186
Sustainable consumption 165, 176-77
Sustainable development 67, 69, 175, 177-78, 251-52
Sustainable harvest 27, 29, 31-32, 35, 38
Sustainable use 29, 32, 59, 71, 83, 159
Synchronous behaviours 203, 206
Synchronous fireflies 203, 205-6
Synchronous flashing 203-5
System
  eukaryotic 103-5, 108

T

Talinum triangulare 223-24
Tamarindus indica 55, 222-23
Target species being 34
Teak plantation 204, 229-30, 233
TEEB exercise in India 159
Testimony of Religious Beliefs 89
The economics of ecosystems and biodiversity (TEEB) 157-58
Therapeutic properties 110
Thorectidae 186
Threatened species 13, 38, 43, 97-98
Thuidium cymbifolium 232-33
Traditional farming systems 84
Traditional healers 61-62
Traditional remedies 27
Trawling nets 199, 201
Tree species shrubs 93
Tree wood 163-64
Tribal concept of biodiversity conservation 56
Tribal people 57, 73
Trigona iridipennis 221
Trimeresurus macrolepis 13, 18
Trimeresurus malabaricus 13, 18
Turtle Excluder Devices (TEDs) 200-201

U

Urinary disorder 80, 85
Utilization
  commercial 27, 32, 35, 252
  sustainable 27, 56-57, 67, 71, 83, 151, 157, 192

V

Validation, scientific 59-60
Value added tax (VAT) 160, 165, 168-70
Vanilla sp. 222, 224
Varieties, farmer's 151-52, 154
Vechur breed 103, 110
Vechur cow milk 103, 105, 110
Vegetable production of India 127, 129, 133
Vinca rosea 222-23
Vitamin B_3 122
Vrikshayurvedic scholars 54-55

W

Water pollution 162-63, 166
Water snakes 4, 11, 145-46
Wealth, country's biodiversity 160
Weedy species 42
  important 146
Wetland ecosystem 235
Whirling beetle 238-39
Wild collection of MAP species 38
Wildlife sanctuaries 56, 67, 97, 249-50

Y

Yeast extract (YE) 241, 243-46

Z

Zingiberaceae 224-25
Zippellianus 232-33
Zooplankton 184-85

www.ingramcontent.com/pod-product-compliance
Lightning Source LLC
LaVergne TN
LVHW070948210926
848482LV00002B/140
* 9 7 8 1 6 8 1 0 8 0 2 2 2 *